FAMINE

BIOSOCIAL SOCIETY SERIES

Series editor: G. A. Harrison

1. FAMINE

Edited by G. A. Harrison

FAMINE

Edited by

G. A. HARRISON

Professor of Biological Anthropology
University of Oxford

OXFORD   NEW YORK   TOKYO
OXFORD UNIVERSITY PRESS
1988

Oxford University Press, Walton Street, Oxford OX2 6DP
Oxford   New York   Toronto
Delhi   Bombay   Calcutta   Madras   Karachi
Petaling Jaya   Singapore   Hong Kong   Tokyo
Nairobi   Dar es Salaam   Cape Town
Melbourne   Auckland
and associated companies in
Berlin   Ibadan

Oxford is a trade mark of Oxford University Press

Published in the United States
by Oxford University Press, New York

© The Biosocial Society, 1988

British Library Cataloguing in Publication Data
Famine.—(Biosocial society series; no. 1).
1. Famines
I. Harrison, G.A. (Geoffrey Ainsworth)
II. Series
363.8   HC79.F3
ISBN 0-19-857662-5

Library of Congress Cataloging in Publication Data
Famine. (Biosocial Society series ; 1)
Contents: Famine | Frances d'Souza—The nutritional
biology of famine | J.P.W. Rivers—The economics of
famine | Meghnad Desai—[etc.].
1. Famines. 2. Food Supply. 3. Starvation.
4. Malnutrition. 5. Autarchy. I. Harrison, G.A.
(Geoffrey Ainsworth), 1927– . II. Series.
HC79.F3F353 1988   363.8   87-31372
ISBN 0-19-857662-5

Set by
Charlesworth Typesetting
Printed in Great Britain by
Biddles Ltd, Guildford & King's Lynn

PREFACE

This is the first of a series of books that the Biosocial Society of Great Britain is planning to produce with the Oxford University Press. The purpose of these books is to examine various contemporary issues and phenomena which clearly have dimensions in both the social and biological sciences and to analyse the interactions and interrelationships between these dimensions. The approach wherever possible will be both theoretical and practical.

As a prelude to publication the invited contributors are asked to present a summary of their position before a meeting of the Society. They then have an opportunity not only of discussing their contribution with co-authors but also of having this evaluated and criticized by a wide audience. It is thought that this process will help to identify gaps and unnecessary overlaps, and 'smooth out' at least some inconsistencies and contradictions. The books should not, however, be thought of as attempting encyclopaedic or even consistent coverage. Essentially they represent a collection of individual views from different disciplinary positions. While these will, we hope, complement one another there will be plenty of room for disagreement, since the problems chosen for discussion are, by their very nature, controversial.

There could not be a more appropriate topic for launching this series than Famine. Depressing in its topicality, it represents one of the saddest practical problems in the modern world. It is also one in which various social, economic, and biological strands and determinants interact in complex ways: fully fitting of a biosocial approach.

By the time famine has arrived there is nothing that academics can offer (except, perhaps, in advising on the logistics of relief, which though heroic often seem to go wrong). But recognized famine takes time to develop, and intervention is at least theoretically possible long before one has a crisis situation on one's hands. Even more important is identifying the various fundamental processes and their interactions, which eventually lead to famine. Only by doing this can the momentum to catastrophe be dispersed at source. Here the academic has much to offer, as this book surely exemplifies.

The four contributors to this book cover a wide spectrum of experience and outlook. Three of them were able to address the Biosocial Society in Oxford on 16 May, 1986.

Dr Frances D'Souza, who has spent much time in dealing with actual famine situations, analyses the circumstances and effects through case-history approaches. Dr John Rivers, who has also had to deal practically with famines, pays particular attention to the nutritional aspects and the bearing these have on the recognition of disaster situations. Professor Desai and Mr Stewart examine the basic causes of famines: Professor Desai with the economic causes, Mr Stewart with the environmental ones. This distinction should not be perceived as an absolute 'either/or' one. Clearly there are complex interactions which vary qualitatively as well as quantitatively from one circumstance to another. Environmental pressures put demands upon economies, and economic processes affect the nature and balance of environments.

This book will of course do little of itself to prevent famine, but we hope that it will help to provide balanced and objective bases from which this tragic phenomenon can be seen and understood, free from both the political and emotional overtones that usually overlie its consideration. And a very wide appreciation of these bases is a prerequisite for prevention and early intervention. What the book certainly highlights is the inestimable value of a biosocial approach to an important world problem.

Oxford   G. A. H.
1987

Acknowledgements: Grateful acknowledgement is made to the Parkes Foundation for a grant which made this book possible.

CONTENTS

CONTRIBUTORS

Frances D'Souza
Department of Biological Anthropology, University of Oxford, 58
Banbury Road, Oxford OX2 6QS; formerly Director, Relief and
Development Institute, 1 Ferdinand Place, London NW1.

J. P. W. Rivers
Department of Nutrition, London School of Hygiene and Tropical
Medicine (University of London), Keppel Street, London WC1E 7HT.

Meghnad Desai
London School of Economics and Political Science, Houghton Street,
London WC2A 2AE.

P. J. Stewart
Pauling Human Sciences Centre, 58 Banbury Road, Oxford OX2 6QS.

INTRODUCTION

G. Ainsworth Harrison

Famine, war, and pestilence seem always to have been an integral part of human history: at least since the beginning of substantial human population growth in the Neolithic. As Thomas Malthus well recognized they have also constantly been regulators of that growth.

There was never much doubt that, worldwide, the twentieth century would be able to do anything about war, but there has surely been a hope that, with developing technology, pestilence and famine would be overcome. Many diseases have in fact been eradicated or controlled, but twentieth-century famines are among the worst in human history, despite the Green Revolution and the surge in agricultural technology. Without doubt pestilence control has been one of the reasons for this, since it promoted further population growth. However, it is far too simplistic to see population growth as the sole cause of famine. The world, as a whole, is able to produce ample food for all its inhabitants and in many areas where people are starving, shops are full of food. Clearly there are numerous interlocking factors that produce famine: many are understood though often nothing is done about them until it is too late; others are only dimly appreciated especially those relating to the perceptions and behaviours of victims and potential victims. Clearly too, there is a need for academic study and wide education.

Perhaps one of the greatest academic requirements, until recently, was to acquire sound, detailed, observational data on just what happens in the development of a famine situation. While the spectacle of the ultimate famine crisis with emaciated dying children in crowded refugee camps has become well enough known, the circumstances that have led up to the full-blown disaster are poorly recorded and little known. This is surprising for, unlike flood or earthquake, famines are not instantaneous events but require months if not years to develop. It is quite possible to document the natural history of famine and work of this kind has at last begun, as is evident from the contribution of Frances D'Souza in this book.

An area where academic study can be of immense practical importance is in developing early warning systems of impending famines and this needs very detailed description and analysis of many environmental, ecological, demographic, and economic circumstances. It is already becoming apparent that different indices are required from one location to another, even in the same country. Having identified what are the most appropriate indicators for a region predisposed to famine then, of course, one needs an adequate monitoring system. This can range from satellite scanning to local observation of crop conditions, nutritional status, and general social integrity. Often enough the practical difficulty is having observers in the right place at the right time, since it is clearly impractical to have a permanent monitoring service at all local levels. Anthropologists might here play a vital role since they tend to be working in areas prone to famine, and by knowing local languages and social organizations are ideally equipped to support interventionist programmes as well as monitoring ones.

It has been assumed so far that a state of 'famine' is easily identified, even if precursor states are not. While the extreme situations experienced in Ethiopia and the Sudan in the 1980s are all too obviously recognized, many people may die of malnutrition or even starvation in circumstances which are not diagnosed as famine. As John Rivers comments, there has been a tendency to define famine through the eye of the observer, particularly the media observer rather than sufferer, but even objectively there is more to famine than just widespread food shortage. In famines practically all social and economic structures break down and the capacity of whole societies to survive is greatly diminished. It is this which prompts the mass migrations which seem to characterize famines and which make famine management so difficult. The background to individual decisions to leave home and all that is familar and face an unknown of great uncertainty – and to do this in a weakened state of body and mind – needs proper investigation. But presumably the 'bottom line to the decision' must be that there is no hope to be found anywhere in staying at home. And that state of mind is as important an element of famine as the food shortage, rampant disease, and social breakdown.

Recognizing famine as a general state provides one form of problem; recognizing those individuals who are most at risk of dying in a famine situation is another. This can be of more than academic interest in famine management, particularly in refugee camps where resources are in short supply and endless life-or-death decisions have to be made by identifying those in most need of food aid. This difficult question is fully addressed by John Rivers in this volume. Less well considered, because it is a little-researched field, is why individuals vary so much in their vulnerability. Even the reasons for the susceptibility of the young and the old are not fully clear, but it is more of a mystery why even within a single family some children are in complete emaciation while others have tolerable body stores. Does this variation represent biological differences or family preferences in food sharing? And in either case 'Why'? The variability in vulnerability, particularly if biologically based, might provide a useful monitoring basis for incipient food shortages since it has been shown that population variance in growth characteristics tends to rise with increasing environmental adversity. This phenomenon is independent of the genetic size characteristics of populations which makes finding common scales of satisfactory growth for pygmies and Nilotes so difficult.

Turning to the causes of famine: as already indicated these are varied and multiple. Generally considered they can be categorized as socio-economic and biophysical but there are important interactions even between these broad categories.

The fact that the worst famines of recent times have been in areas torn by civil strife, such as Northern Ethiopia, the Ogaden, and parts of Mozambique, is not coincidental. It is in the chaos of war that social and economic breakdown is most likely and most damaging. It is also of course under such circumstances that it is most difficult to provide and organize aid. However famines have by no means been confined to politically unstable areas. Economic factors in their own right can play a crucial role as is well demonstrated by Meghnad Desai in this volume. It might have been thought that peasants and farmers would usually be the best protected against food shortage because of their proximity to agricultural production. In a fully traditional subsistence economy this is the case, but societies with these economies do not experience 'famine' probably because the 'scale' of any food shortage is so small and local that it is never designated so. And as soon as one moves to more advanced economies, agriculture is primarily concerned with cash cropping. It is factors which produce extreme poverty, and do so on a large scale which cause famine. And in this situation it is the peasants who are the most vulnerable; they are the ones who are least likely to have money to buy food even when it is present in shops.

The seminal role of economic factors in famine causation have become widely recognized since the pioneering studies of A.K.Sen. This should not however be taken to imply that environmental factors are of lesser importance. Twenty-one countries in Africa experienced severe drought during 1984–85, and this as much as anything was responsible for the widespread famine over the continent at that time. Various climatic factors especially rainfall, can affect agriculture and the quality of the harvest as can other physical elements like soil conditions and biotic factors like pests and crop and animal diseases. These are the considerations of Philip Stewart in this book. All the contributors recognize, however, the complex interactions and chain of events which lead to famines. For example, social, political, and medical changes can lead to overpopulation and this to overstocking and overgrazing. These promote soil erosion and climatic changes leading to drought over a wide area. Crops fail and especially vulnerable are monocultures. If such monocultures are of cash crops, as they often are, peasants receive no money and are unable to buy the food available. Clearly one has the makings of an extensive famine, particularly if through political instability and war, the dangers are not recognized as sufficiently important to seek advanced help. How ironic that in attempting to produce food for consumption elsewhere peasants starve because they cannot afford the food produced elsewhere!

Some uncertainty in food supply has long been part of human experience. It arises through climatic seasonality if nothing else. And it would appear that human beings typically have inbuilt mechanisms, at both the biological and cultural level, to cope with the vagaries that are part of every environment. But the modern world has generated economic and biophysical circumstances which are far beyond anything in past experience, and which cause environmental and socio-political fluctuations which are new, in degree and in nature. To deal with these co-operation at all levels, from research and development to financial and political aid, is needed. This will only come when it is properly appreciated that a famine in some part of the distant world is not just 'their problem' it is 'our problem', for we surely have played our part in causing it.

1

FAMINE: SOCIAL SECURITY AND AN ANALYSIS OF VULNERABILITY

Frances D'Souza

INTRODUCTION

THE NEED FOR RESEARCH

Starvation in Ethiopia took the world seemingly by surprise in the autumn of 1984. By 1986, apart from perennial pockets of drought in Sahelian West Africa, the rains were falling and drought-stricken Africa rapidly became green. Widespread famine is over for the time being, the media can go home and the International Donor Community (including bilateral, multilateral, and non-governmental donors, hereafter referred to as the IDC) can return to deliberations on development, loans, and the potential for industrialization in developing countries. Perhaps up to a million have died in the Eastern and Southern Africa famines between 1983–5, some few on camera. The nightly prime-time television pictures of starvation, of food being transported by the hundreds of tonnes or languishing in sky-scraper mounds on docksides, have given way to other, more current, disasters. The slow, but crucial, business of rebuilding communities ravaged by famine is not mediagenic unless at the same time a major scandal can be unearthed. It is time now for the researchers to move in and go over, in some detail, the harrowing events which preceded and accompanied mass starvation and to plead once again that the world be better prepared the next time round.

It may, however, seem a thankless task since the facts which have emerged so far indicate that in Ethiopia – as in the Sudan, the Sahel, and Mozambique – famine had clear antecedents which were either inade-quately recognized and communicated, or simply ignored by donors and the governments of victim countries. It is, perhaps, the duty of the research community to reconstruct the story from its beginnings in order that eventually a better understanding of impending famine and its complexity can emerge and be acted upon. Has anything been learnt by researchers or administrators? Have any lessons been documented for future reference? Can systematic information, albeit retrospective, guide decision-makers to act more appropriately not only in the future but also today in helping to reduce vulnerability in famine-prone areas of the world? The answer to the last question must be yes, in the long term. Famine is as much open to scrutiny and analysis as any other social phenomenon. Furthermore, increased and more widespread understanding of famine can only help to promote more timely and appropriate preventive action. It is, therefore, the overt theme of this chapter that the apparently arid task of reconstructing human tragedy is an important, indeed vital, task. The attempt here, therefore, will be to describe and analyse those factors in developing countries which interfere with and disrupt those fundamental social and economic activities which in themselves reduce vulnerability to famine. The hypothesis to be tested can be stated very simply: what are the necessary and sufficient conditions for the emergence of famine? This hypothesis will become increasingly refined as recent research is discussed, but essentially the aim will be to understand and present those factors or events which *prevent* individual, household, or community access to food and, therefore, what inputs from the aid community can best help to *restore* or *maintain* access to food.

Logically the answer to these latter points has three requirements: an understanding of what *causes* vulnerability to famine; of what *precipitates* starvation; and of what *characterizes* the beginning of mass starvation. Later in this chapter these issues will be discussed with reference to recent food emergencies in Lesotho (1983–5), Mozambique (1982–5), and Ethiopia (1983–5). These case examples each present very different precipitating factors as well as very different responses on the part of the IDC. A clearer picture of what appears to motivate donors to act, and how these motivations might be modified in the light of actual patterns of giving and not giving, can be drawn.

Broadly speaking one can distil two world-views on famine from a considerable academic and other literature. *Firstly*, that famine, in the twentieth century, is an insult and need never happen in a world abundant with food. *Secondly*, the less sanguine view is that famine is inevitable due to population growth and consequent increased demand on scarce and uncertain resources. In fact neither view is tenable since they both assume truths about the nature of famine which are not borne out by facts. Nevertheless this chapter questions the second view and considers the kind of commitment required from both the IDC and developing-country governments to allow the former view, if not to prevail, then at least a fair hearing. The rationale is twofold:

1. If the exact cause of famine in a given community can be defined and the precipitating factors identified, then, theoretically at least, famine

can be prevented. This must be the case if famine occurs, as it does, in the context of overall sufficiency and in a world where the Western-aid machinery is publicly committed to eradicating this biblical scourge. The fact that famine is apparently unsuccessfully alleviated let alone prevented must, therefore, be due in part to a lack of information on causes and indicators resulting in a late and, as far as saving lives is concerned, ineffectual response.

2. It is also the case that the combined resources of the IDC are massive and can swell to staggering proportions in a well-publicized emergency. This resource represents considerable power and influence. This chapter will also seek to understand how such power is currently perceived and used and how its use might be improved to establish solid famine prevention mechanisms in vulnerable communities of the developing world.

But it can be argued that even were there perfect understanding of the famine phenomenon and concerted action by people of goodwill, might there not remain insurmountable barriers, given the state of underdevelopment in famine-prone countries and the political motivations of major bilateral donors? Perhaps, but these views tend to assume weight and respectability without being rigorously examined. As an exercise let us look at three commonly quoted obstacles to famine prevention and relief.

   (i). *The Information Gap*: That there is a scarcity of useful and systematic information from the field is not in doubt. However, experience of recent African famines (let alone those which have occurred in South Asia) suggests two things. Firstly, that a great deal of information – of variable quality admittedly – does in fact exist; and secondly, that it is extremely rarely used by donors or by governments. Part of the reason for this is that it is essentially unusable; either the methods of collection themselves appear to be so arbitrary that the results are questionable, or there is no single authoritative body capable of collecting, interpreting, and issuing relevant information to operational agencies. A vicious circle exists here as in other stages of the famine cycle; donors are persuaded to act when they are sure (i.e. have irrefutable evidence) that a famine crisis is upon them. At this stage information collection as to who exactly is starving, where they might be, and how they might be reached is often considered a luxury when the pressing need is to be seen to be getting food on the road. But the pressure to collect and use *pre-famine* information is absent precisely because there is no 'emergency'.

   Preliminary estimates indicate that the expense of setting up a professional information-collection network devoted to monitoring vulnerability to famine is prohibitive. Yet without it there is only a slim chance that impending famine can be regularly detected and prevented. As always there is a compromise solution. This is to persuade field-based agency representatives, consultants, anthropologists, and anyone else with regular access to vulnerable communities to collect simple but indicative information. Equally, and perhaps even more important, it is necessary to ensure that the increasing number of surveys carried out are done so with a degree of professionalism. Above all, many lives could probably be saved if the authors of such surveys and reports were prepared to share them with the donor community. This should not be a wholly unrealistic expectation, but as long as it remains unfulfilled little real importance and priority will be given to information collection, and the anecdotal reports of starving peoples traversing a country in search of food will continue to be dismissed as alarmist. The information gap, therefore, is a crucial weakness but neither structural nor insurmountable (see, for example, D'Souza 1984b, 1985a; D'Souza and Shoham 1985a; D'Souza 1986b).

(ii). *Politicization of Aid*: Aid, it is said time and again, is given, even in the context of an emergency, to satisfy the current and explicit political designs of major bilateral donors. The implication is that nothing can be done about it.

In this view, the world and its inhabitants are pawns in a larger and more deadly game, and any attempt to alter the relentless march of competing powers to control the destiny of developing countries is naïve and therefore futile.

It should be said at the outset that increasing politicization of aid is undoubtedly a factor in famine relief but it is not the whole story. Many financially powerful agencies genuinely seek to retain control of their own resources and to meet need regardless of where those needs occur. More importantly, however, the research community in this particular field of endeavour cannot passively accept this view. The responsibility and the task is, first of all, to understand better how communities become vulnerable, then to deduce as a result what kind of action at what stage (before, during, and after famine) donor agencies can take to minimize suffering and death. The act of conducting this kind of research and publishing it, is in itself a constructive means of controlling the politicization of aid.

At present, there are no clear examples of a deliberate and callous refusal to act due to political considerations. This is particularly true of the recent Ethiopian famine even though it was preceded by considerable predictive information. The information was confusing and it is hard to claim that it provided a sufficient basis for action. If, however, the research community had been better equipped at that time to document precisely the, by now more acceptable, indicators of forthcoming and widespread distress, and had that evidence been forcibly and publicly presented as a joint effort by agencies represented in Ethiopia, then donors might have been persuaded to act earlier. At least they would have been less able to procrastinate.

It will probably be more difficult for donor governments to plead ignorance in the future due to an increasingly sophisticated media and a greater concern and understanding on the part of the donor public of famine forewarning. Nevertheless, the research community can significantly speed up this process by continuing to refine the indicators of famine. One can thus argue that, at the present time, a late response can be more likely attributed to lack of precise and usable information rather than purely to evidence of political manipulation.

(iii). *Logistics*: Historical records are full of examples of famine resulting from the lack of transport to convey food from surplus to deficit areas. For example, during the later stages of the Roman Empire, while sea routes were highly evolved, inland communications remained primitive and food could not be transported to starving villages of the interior (see Dando 1980). Today the situation is little changed in many famine-prone countries, especially in Africa. While logistic costs are always high and budgets difficult to agree and allocate, the persistent underlying difficulty lies in the scarcity of trucks, spare parts, fuel, trained mechanics, and, above all, in a basic system of management. This may not seem immediately relevant in a review of famine and its wider consequences, but many hundreds of lives are lost in food emergencies because transport is grossly inadequate or, simply, non-existent.

The point of raising this fundamental weakness at this stage is to point out that of the many millions of dollars available once an emergency has been declared it is surely not impossible to set aside at the outset the necessary finance and expertise to deal with this monotonously regular obstacle.

A deliberately optimistic tenor to this chapter has been set – famine is preventable – more so than any other kind of disaster precisely because it has antecedents, a warning period during which planning can take place. However, famine prevention must also go beyond ensuring the rapid provision of free food to communities at an early stage. The aim must be to direct development aid to bolstering famine-prone communities, in order to reduce their vulnerability. To do this one has to develop a very detailed understanding of what causes vulnerability in different social and economic contexts. There is no better starting-point than summarizing what recent research has to say about famine, which is surely a manifestation of the most extreme vulnerability.

THE PHENOMENON OF FAMINE

Famine is a distinct economic and social phenomenon. It differs from endemic and even severe malnutrition due to poverty. It can be characterized by its profound and long-term disruptive effects on communities. Famine, as is the case in other, so-called 'natural', disasters is unquestionably a social event – and one can broadly describe the phenomenon as a series of increasingly desperate actions by people to obtain food (see Seaman and Holt 1980; Sen 1981).

Even the extremely poor achieve a balance with the society in which they live. They may not have access to health care, clean water, adequate shelter, or regular income, but there is usually a system whereby food, often poor in quality and quantity, can be obtained and the social unit, however it may be defined, functions to ensure bare survival. Of course, the balance is extremely fragile and can be destroyed even by small shifts in the external environment. For example, pavement squatter groups in South-Asian cities can die of starvation if suddenly forced to move to an area in which no relationships of debt, obligation, or work have been forged, and insufficient time or too great competition exists to create such contacts.

The immediate impact of errors in the system then will depend on what reserves the individual holds. These reserves may be cash, food, or relationships which can be activated in order to obtain food. Different degrees of vulnerability to acute food shortages will, to some extent, depend on the local economy. For example, the rural subsistence farmer may be even more vulnerable than the urban squatter. Although he may have direct access to food he himself produces, he may also have less opportunity to diversify his income sources or, indeed, to spread the risk of his investments. Thus in a bad season or series of bad seasons, his food reserves may become so depleted, that in the absence of other resources such as manpower to offer in return for food, his family's position becomes vulnerable. In these circumstances the existence of strong social networks, which carry obligations to feed even remote relatives, becomes an important protection. Similarly, in a situation where food prices rise exponentially a wage labourer can be highly vulnerable in circumstances where a sharecropper is better protected (see Sen 1981). None of these individual and tragic events, however, defines famine though they may increasingly occur in pre-famine conditions. Rather famine is characterized by abnormal social and economic responses to threatened or actual food shortages. Typically, these responses tend to make people even more vulnerable, especially for the future, and thus famine has its own momentum, regardless of weather patterns. The implicit feature of pre-famine and actual famine is that whole communities tend to become involved in such abnormal responses. Starvation, on the other hand, which is the culmination of a number of discrete events, varies due to physiological resistance of the individual and, for a period at least, different degrees of access to remaining but increasingly scarce resources. In the absence of any intervention at this stage, starvation affects first the most vulnerable such as the young, the old, and pregnant and lactating women, and, eventually, all members of the community. Death itself is often precipitated by infectious or contagious disease (see Seaman and Holt 1980).

Famine can, therefore, be distinguished from starvation though the one more usually leads to the other. The implications of this distinction are several, but for the purposes of this chapter the bearing it has on understanding the indicators of famine is of great importance. For example, if one asserts that famine in its early stages is characterized by social and economic responses, then it cannot be detected by anthropometric or nutritional means, or even defined in these terms. Significant changes in body weights and measurements of children as a direct result of lack of food occur relatively late in famine, and to rely on this kind of evidence inevitably means that relief will arrive too late (see Borton and Shoham, 1985).

Failure of a given crop or even several crops in consecutive seasons need not result in famine if there are other resources in the community, or access to resources outside the community. War, even if it is prolonged, does not necessarily cause famine if the opportunity to plant and harvest crops or to obtain food by other means is present. The remarkable ability of peasants to avoid famine and widespread starvation in Cambodia between 1975–9 was due in part to the traditional fertility of that country. By contrast, the Khmer refugees who streamed across the border into Thailand were starving, and this may have been due to the fact that the social networks which regulate the production and distribution of food had completely broken down (see, for example, Murlis 1980; Shawcross 1984).

To summarize then, famine can be defined as a reduction in normally available food supply such that individuals, families, and eventually whole communities are forced to take up abnormal social and economic activities in order to ensure food. If these activities are unsuccessful, then starvation will follow.

FAMINE OBSERVED

But what are these abnormal activities? It is now fairly generally accepted that famine is rarely, if ever, the result of absolute food shortages but due rather to the unequal distribution of available food. If the normal supply of food for one reason or another is restricted, there may follow sudden and sharp price rises with the result that poorer people simply cannot afford to purchase it. Price rises are frequently due to threatened rather than actual food shortages as, for instance, typically occur when rains destroy part of a standing crop and market-traders anticipate a reduced supply. Whatever the cause of such speculation and panic, sudden price rises prompt a whole series of further but none the less abnormal actions on the part of the peasant aimed at maintaining his purchasing power. The following responses have been observed in countries as geographically diverse as Ethiopia (Seaman *et al.* 1974, 1978; Seaman and Holt 1980; Sen 1981; Cutler 1984; Holt and Cutler 1984); the Sudan (Tobert 1985; York 1985; Cutler and Shoham 1985); India and Bangladesh (see Sen (1981) for full bibliography on a considerable literature); Afghanistan (Fry 1974; D'Souza 1984a); Mozambique (Unicef 1984, 1985; D'Souza 1986b); Europe (Dando 1980; Smith and Christian 1984), and China (Snow 1962) though not necessarily in the same sequence, or with the same intensity or spread.

As food shortages continue, for whatever reason, and prices of an increasingly valuable resource rise, individuals are driven to raise cash to buy food for survival. The interval between price rises and first signs of starvation can be extraordinarily variable, and appear to be dictated by the degree to which there are reserves in a given community. Thus price rises may be gradual as consumers use their own reserves, call in debts, or are protected by relatives; or they may be quite sudden. A 58 per cent rise within a four-week period has been recorded for markets in Rangpur, Bangladesh (see Sen 1981, p. 149) and was apparently due to sudden and unprecedented demand on the market, combined with hoarding and speculation by marketeers. The farmer is forced to sell what assets he has in order to raise cash for food. These assets may be in the form of labour and it is common in the early stages of food shortage for younger males of a community to move to urban areas or even across international boundaries seeking work and wages. In urban areas there are reports of increased prostitution and incidents of theft (see D'Souza 1986). Less valuable assets such as sheep and goats may then be hastily sold and subsequently, increasingly valuable assets such as cows and plough oxen, leaving the individual farmer virtually destitute when the crisis is over. As animals flood the market the tendency is for livestock prices to drop and, in the absence of any external intervention, the farmer's purchasing power is steadily eroded. Meanwhile, there are consistent reports of consumption and even sale of so-called 'famine foods'; these are naturally occurring plants, fruits, or berries which are not normally consumed, let alone sold. At one stage of acute food shortage in rural Mozambique the local market was selling *only* famine foods (see Unicef 1985; D'Souza 1986b). The great famines of England in the early Middle Ages, and in Russia and China, document the sale of children and, more horrifyingly, cannibalism (see Dando 1980; Smith and Christian 1984).

As a last and desperate measure the farmer and his household leave their settlements to trek to food distribution centres or to where there is even the rumour of food. As has been repeatedly recorded and documented, this 'distress migration' (see, for example, Cutler 1984) is all too often perceived as the first sign of serious crisis by the IDC. In fact, it is a terminal indicator and at this stage it is almost always too late, for logistic reasons alone, successfully to avert massive starvation.

However, a description of the events which appear consistently to precede and accompany famine does not necessarily further an understanding as to what causes famine. For this one has to look to recent research which attempts to incorporate such observations into a more general causative framework. Much of this work (Seaman and Holt 1980; Sen 1981) challenges the outdated, but still held, view that there is a necessary and direct relationship between drought, crop failure, and famine. This is in spite of an historical record replete with examples illustrating that famine is rarely, if ever, a natural disaster, but almost always occurs in contexts of upheaval and war, or because of massive differentiation between the rich and the poor whereby the latter are precluded from access to the food that they grow.

J. Penkethman, writing in the mid-seventeenth century, attributed the famines which occurred in England between 1066 and 1638 to the wars which wasted both corn and land, to the weather, and to what he referred to as the 'abasing of the coin' and 'the uncharitable greediness and unconscionable hoarding of corn master and farmer' (quoted in Dando 1980, p. 122). Pierre Monet (1964) records contemporary observations of early Egyptian famines:

The Nile cannot be held solely responsible for the great famines which wrought havoc in Egypt on more than one occasion. The disturbances which became fairly general during the First Intermediate Period and even more so under the last of Ramessides, meant that canals and roads were left untended, trade and building operations suspended and the cultivation of crops brought to a standstill. There was little incentive to sow crops if bandits were likely to make off with the harvest. During the last years of the Twentieth Dynasty, the price of food continued to rise and people pillaged temples and tombs in order to obtain gold (Monet 1964, p. 77).

Finally, records of the devastating famines in China and Russia in the last 200 years indicate that they are generally 'considered as visible manifestations of deep rooted internal conflicts and exploitative systems' (Dando 1980, p. 84). Weather and consequent crop failure is not, and never has been, the only or even the main cause of famine. Famine is not a natural disaster.

Sen, in his book *Poverty and Famines* (1981), critically examines and then dismisses what he calls the 'Food Availability Decline' approach because such an approach says little about who is dying, where they are dying, and why, and instead concentrates on the causal mechanisms. These, he concludes, may be more concerned with a breakdown in the individual's entitlement to food. Thus the observed unequal impact of drought results from individuals' differential ability to 'command food and other essential goods' (p. 153). A person will be exposed to starvation if what he rightfully has to offer does not buy him sufficient food. At this point only 'non-entitlement' transfers (i.e. charity) can save him from starvation.

Therefore, although availability of food by no means precludes famine, sudden changes in people's *access* to food virtually guarantees famine and even starvation for a given sector of a community, in the absence of any kind of social insurance or external assistance. These changes may be due to job loss, cash depreciation and/or exponential price rises, enforced migration, or war causing displacement. Sen implies that famine is a result of the breakdown of social and economic contracts that an individual has formerly held with owners or controllers of food supplies. In this sense it can be said that famine is characterized by social and economic disruption.

This has clear implications for both the detection of impending famine and its prevention. 'What is needed is not ensuring food availability, but guaranteeing food entitlement' (p. 129). But to whom should food entitlement be guaranteed and when? The answer will depend on a thorough understanding of which is the most vulnerable sector in different communities and what has caused that vulnerability. Entitlement, as Sen points out (and, therefore, potential vulnerability), depends on what an individual 'owns, what exchange possibilities are offered to him, what is given to him free, and what is taken away from him' (pp. 154–5). An analysis of these assets and opportunities would sort out who would be most likely to survive even consecutive seasons of drought and who would be most likely to succumb in the first drought season.

Genuine famine prevention in the future, therefore, may be far more concerned with understanding the cause and incidence of failures in entitlement and in ensuring through development inputs that individuals maintain, at all costs, that entitlement, rather than in the provision of free food just prior to or during starvation. While it is clearly vital to continue attempting to relieve starvation, it is nevertheless an admission of failure in development and, therefore, also in famine prevention.

The next section of this chapter will look at general vulnerability as a consequence of underdevelopment but also at specific vulnerability and its causes. The hypothesis given in an earlier section of this chapter can now be more closely defined: If famine and starvation result from a gradual or sudden decline in the individual's ability to purchase or otherwise obtain food then those factors which interfere with normal purchasing power such as social disruption as a result of war, migrant labour schemes, and relocation policies, must be included as causal factors since they radically undermine the individual's or the household's capacity to withstand famine. The task, then, is to understand precisely how and why coping-strategies break down.

CASE STUDIES

LESOTHO, 1983-5

The Emergency

In March 1983, following a substantial reduction in seasonal rains the Government of Lesotho (GOL) declared a drought emergency and in June 1983 issued an international appeal for nearly 120 000 tonnes of food to offset the projected food-gap. This request was later modified to 31 000 tonnes after an agreement to increase commercial imports for the non-priority drought areas. The donor community, for the large part, responded rapidly and generously leaving a final estimated food requirement of just over 12 000 tonnes.

The evidence put forward by the GOL for the emergency was based on the observation that the rains between November 1982 and January 1983, essential for the planting and growth of the 1983 crop, had been reduced in some areas by a third or even one-half of a nine-year average. Should the drought continue, the GOL predicted a 50–70 per cent crop loss and substantial livestock mortality. By June 1983 the GOL reported that the drought was proving to be as severe as that of 1933 and that grain-crop failure was estimated to be between 70 and 80 per cent.

Anecdotal rather than survey reports from aid agency staff as well as from government sources indicated that the poorer mountain communities were the hardest hit and that the imminent winter would cause the severest hardship. The total number reported to be in need of emergency food rations was 501 346 (i.e. 43 per cent of the *de facto* population) and by September 1985 a minimum of 415 000 had, in fact, received rations of one kind or another.

The criteria for inclusion in the relief programme consisted of a positive reply to any three of the following four questions asked of household heads; if there were:

- insufficient harvest and/or food to last beyond five months
- insufficient money to purchase food beyond five months
- less than 35 head of sheep, goats, or cattle
- a disabled person as household head or one in which the head was a widow.

No account was taken of current access to regular food aid programmes – in themselves considerable and serving perhaps up to 50 per cent of the total *de facto* population (see D'Souza and Shoham 1985b). Other than clear evidence of substantial failure in rains and the fact that extreme poverty exists, especially in mountain areas, there was no indication of famine or starvation. The questions that arise include: what impact would one expect drought to have in Lesotho, and, depending on the conclusion, what kind of assistance was called for? A short history of the declining Lesotho economy and, in particular, the local coping-strategies which have evolved, will clarify both these questions.

Economic Background

National Vulnerability

There are several factors, economic, geographical, and political which combine to make Lesotho potentially vulnerable to serious food shortages. The country is small, completely surrounded by the Republic of South Africa (see Fig. 1) upon which it is economically dependent and, in spite of a growing population, the 13 per cent of land available for cultivation is decreasing, some say at the rate of 2 per cent a year due to over-use and erosion. Over 70 per cent of the population is concentrated in the narrow

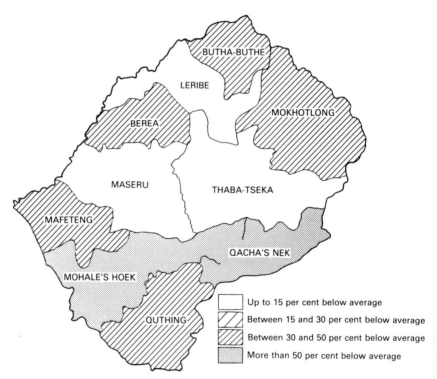

Fig. 1.1. Lesotho: production of grain crops in 1983–4 compared with the 10-year average.

truly lowland corridor of the west. The mountain regions of the north, west, and south support small-scale subsistence farming, and during the winter months these communities may be entirely cut off from trading and economic centres. Communications networks are very poorly developed and, thus, the cost of importing food to mountain regions is high.

Lesotho has never been economically self-sufficient and its trade imbalance has increased every year since independence. The three factors which support Lesotho's viability are: membership of the South African Customs Union (SACU) which has provided approximately 70 per cent of the GOL's total revenue in recent years; the opportunity for up to 200 000 of the economically active force to earn substantial wages in the RSA as miners, farm labourers, and domestic servants; and finally, a generous aid response from Western donors which, for example, in 1984 provided 31·4 per cent of the country's total food requirements through emergency donations (Table 1.1).

TABLE 1.1. Lesotho: country statistics, 1984–5

1. *Population*	1·36 million
Average density	45/km^2
Range	40–505/km^2
Mortality rate	16–18/1000
Birth rate	38–40/1000
Growth rate	2·6%
Literacy rate	60%
2. *Land*	
Area	30 355 km^2
Percentage cultivable	13%
Main crops	Maize, wheat, sorghum
Livestock (estimated)	2·7 million
3. *Economy*	
GDP	US$259 million
GNP/capita	US$380
No. of registered migrant workers	158 000★
as a percentage of economically active population	23%
Remittances	US$38·2 million
as % of GNP	40%
Cereal imports 1984/5	200 000 tonnes
Development aid	US$62·3 million
as % of total revenue	35·7%

★ This figure represents registered workers in gold-mines. Other, unregistered mine-workers and seasonal labourers bring the figure nearer to 200 000.

Sources: Sub-Saharan Africa Year Book (1985). Europa; *Annual Statistical Bulletin* (1985). Government of Lesotho, 1985; *Information Bulletins*, **1–4**. Food and Nutrition Co-ordinating Office, Government of Lesotho, 1985.

The relationship between declining production and increased remittances is shown in Table 1.2. Wage increases negotiated by unions have ensured a high level of revenue for Lesotho as a whole and it is probable that the majority of families have at least one male relative, albeit a remote relative, earning a wage in the Republic.

The opportunity to earn wages is vital for agricultural Basotho families but the heavy reliance on this income source is subject to political decisions made in the Republic. Should the RSA decide to limit future enrolment, or worse, repatriate existing mine-workers it would be a disaster for Lesotho of a magnitude difficult to comprehend and certainly almost impossible for the aid community to mitigate in the short term.

There is in Lesotho, as in any developing country, a marked variation in access to resources and a differential vulnerability. Remittance is the basis for economic activity. There are some who, unable to migrate, without land and livestock and with little or no education, cannot earn remittances and may therefore find themselves to be absolutely poor (Spiegel 1979). This basic vulnerability is caused and exacerbated by several facts operating at both country and household level. A high population growth rate and steadily declining investment in agriculture means that the 30 000 or so who annually enter the labour market will be landless. Employment within Lesotho is negligible, and thus the individual is forced to migrate in order to survive.

Improving agriculture and production would be a Herculean task given the inauspicious climate of extreme cold, scorching sun, hail, frost, and torrential rains, and the fundamental lack of investment in seeds, fertiliser,

TABLE 1.2. Estimated production figures for main crops and remittance and deferred payments, 1973–83

Year	Maize (tonnes)	Wheat (tonnes)	Sorghum (tonnes)	Deferred payments	Remittances (Millions of Maloti)
1972/3	70 000	36 000	43 000	4·6	3·9
1973/4	122 500	57 000	84 000	7·3	5·1
1974/5	70 292	45 337	37 443	12·8	7·1
1975/6	49 128	44 640	24 540	17·8	8·2
1976/7	125 932	61 381	62 313	16·6	10·9
1977/8	143 168	57 906	85 775	20·3	12·9
1978/9	124 856	33 629	69 952	22·6	15·4
1979/80	105 618	28 194	59 285	24·4	17·6
1980/1	105 674	16 993	47 729	35·8	26·9
1981/2	83 028	14 462	26 158	76·7	51·0
1982/3	76 200	14 810	30 687	103·5	74·2

Source: Annual Statistical Bulletin (1983) Government of Lesotho, March 1985.

and any degree of mechanization. This, however, was not always the case and Lesotho had its Golden Age in the 1860s when it was virtually a granary for the Orange Free State and parts of the Cape Colony. In addition Lesotho supplied food to the fast-growing mining communities in Kimberley. During the late nineteenth century the economy became highly monetized and commercialized. For example, taxes although theoretically possible to be paid in kind, were in fact demanded in cash. A gradual and then rapid decline occurred in the subsequent 50 years due to the apportioning of fertile land to South Africa, to soil exhaustion, population growth, and factors such as the development of transport networks which entirely bypassed Lesotho (see Wallman 1969; Spiegel 1979; Murray 1983). Today Lesotho is an impoverished labour reserve for the RSA and even basic foodstuffs are imported. Cobbe (1982) gives an illustrative example of frozen chickens being available in surprisingly remote rural areas because, since farmers produce no surplus grain, the cost of purchasing a frozen chicken was less than the transport costs of importing chicken feed!

Household Vulnerability

The fact of migration and the economic imperatives that sustain it have given rise to many coping-strategies and the genesis of vulnerability and poverty can be seen when these strategies break down. Several attempts have been made to categorize relative wealth and poverty and to illustrate coping-responses with the use of household case studies (see Wallman 1969; Barker 1973; van der Wiel 1977; Spiegel 1979; Gay 1984; Bryson 1984). The following examples are abstracted from this literature in an effort to summarize what constitutes vulnerability and how some families manage to escape from it.

A woman is abandoned by her husband who has meanwhile remarried in South Africa and who never completed bride-wealth payments for his first wife. In the absence of remittance money she leases her former husband's land to his relatives with an agreement that she and her children have a share in the produce. In the following season the crop is partly destroyed by drought and her share is insufficient for survival. Her own family claims her economically viable children in lieu of defaulted bride-wealth payments and she is forced to seek employment wherever she can. Her survival will depend on there being an adequate crop in the next season, her capacity to earn an income, remarriage, or receiving some bride-wealth payments when her remaining daughters get married. The important event in her life which almost guaranteed destitution was the lack of cash to sustain her fields.

The second example concerns an old man, a member of an extended family who has leased some of his fields to junior members but continues to

plough those which remain himself. He is also caretaker of small livestock
and has the right to sell their fleece; he is preferentially hired to work on
the fields of the livestock owners. He occasionally receives small gifts from
returned migrant family-members and he owns two donkeys which he hires
out as pack animals to transport maize earned by other men on food-for-
work projects. His fourth and much younger wife occasionally crosses the
border to work as a domestic in South Africa.

The third case example illustrates relative wealth. A larger than average
extended family owns land and livestock and continues to receive remittances with which trading activities as well as profitable farming are subsidized. The family owns and works a café alongside the main road and the
wives brew the beer sold. Wives also occasionally act as seamstresses using
a sewing-machine bought with remittance money. From time to time male
members of the family, in between mining-contracts, will work in relatives'
or neighbours' fields receiving cash or food as payment.

These case examples illustrate, if nothing else, that farming is heavily
subsidized by remittance money and, therefore, is not on its own a viable
source of income. The fundamental dilemma for the rural household is that
'there are too many people for the available land and resources, but too few
to provide an adequate labour-force properly to exploit these resources'
(Wallman 1969, p. 45). Livestock is an important, although not highly
profitable, source of cash and for many Basotho it is the principal, if not the
only, source of income. Furthermore, bride-wealth though traditionally
paid in cattle is, today, the exception rather than the rule. The fact that few
if any can meet this traditional requirement means that the obligations
incurred by proper bride-wealth payments may not operate and this can
cause the poor to become destitute. A man unable to pay more than a small
part of the bride-wealth agreed loses the rights to his children. Therefore
in the event of domestic misfortune he cannot claim access to their services
and eventual income because of the lien upon them claimed by his wife's
kin.

The balance in many households between coping and near destitution is
thinly held and domestic tragedies, especially those which interfere with
marriage, are deeply destructive. As is clear, the viability of the household
economy requires prolonged separation which in turn seriously jeopardizes
the stability of marriage. The larger family with more diverse sources of
income is apparently better protected than the small family which relies on
a single and uncertain source of income. Smaller households are also more
dependent on mechanisms within the community for the transfer of
income between households. But in spite of the fact that the larger family is
more viable the indications are that there is an increasing trend, in the last
two decades or so, towards nuclear families.

The average landholding is 2 ha, but even this requires a substantial labour and cash input to ensure a reasonable yield. Ploughing, planting, weeding, and harvesting require the use of a tractor, money for seeds, and possibly for fertilizer, and intensive labour to ensure that the maize is kept free of weeds during the growing season. If cash is available these goods and services can be bought. If not, then other options include sharecropping arrangements which are often fraught with difficulties, and the arrangement of work parties whereby labourers are guaranteed food and beer – the latter item presupposes that the household has sufficient surplus sorghum to have brewed beer.

Raising extra cash can be achieved through petty trading, pig breeding (a popular activity if the initial outlay can be capitalized, because pigs live off the residue of beer brewing), sale of livestock, and/or prostitution. If the farming household still cannot obtain sufficient food through these activities they may decide to double crop which not only exhausts the soil further, and, therefore, reduces yields, but also interferes with the traditional and highly valued practice of allowing village livestock to graze the stubble after harvest. Part of a harvest may be sold to raise cash and later bought back from traders at much higher prices, representing a net loss in investment. As a last resort, a woman household head who cannot raise cash to farm, may cross the border to work illegally in South Africa at grossly exploitative rates and moreover risking prison. Her children meanwhile will have to be lodged with family or neighbours and she may lose the right to their resources if and when she returns. Studies of poverty (Feacham *et al.* 1978; Spiegel 1979; Murray 1983) indicate that the poorest households are those in which the permanent household head is a woman, implying no regular cash remittances, or those which are small, i.e. having four or less members. This is explained by both the kind and amount of investment that agriculture in Lesotho requires. One of the most striking features of rural Basotho families is the imbalance between the sexes resulting in a skewed *de facto* population of the old, the young, and women. It is quite normal for up to 50 per cent of the young to middle-aged male population of a given community to be absent more or less permanently and in some studies over 60 per cent absenteeism has been reported. The effects of these enforced separations are many and varied not the least being the threat to marriages of prolonged separation. Women are now the effective managers of households and farms. Social and economic co-operation in recent years has, therefore, become pragmatic rather than traditional. For example, Wallman (1969) reports that the tradition of hospitality has changed and 'nowadays not even the verbal pretence of such hospitality is kept up; villagers will stress instead their own poverty and the perfidy of strangers' (p. 57). Severe strains are also introduced into family and kin relations which are by no means resolved in spite of perhaps three to four generations of a migrant way of life. Conflicts frequently arise over priorities in spending remittance money, management of farm land, leasing arrangements, and the legitimacy of children.

It can be said that vulnerability is both caused by and a result of social disruption due to enforced migration. It is generally agreed by students of Lesotho that agriculture seldom provides even minimal security to the household without regular access to cash. The prevalence of cash in the economy over a long period has allowed people to survive but has also increased vulnerability because of the inherent problems of relying on a resource *outside* one's own political boundaries. The tradition of migrant labour may also have encouraged only a limited investment by the government in development schemes such as irrigation projects, cash cropping and other income generating-activities. Lesotho's underdevelopment can thus be seen partly as a result of the long-existing access to cash in the RSA. The question, in this context, is how vulnerable is this kind of an economy to famine and what effects might substantial amounts of free food have on the local economy in the event that there was no famine.

The Relief Response

There is no doubt that famine of the kind seen in Ethiopia, East and West Sudan, and some Sahelian countries did not occur, and probably never has occurred, in Lesotho. The classic signs of impending and actual famine – unusual migration, price rises in staple foods, increased sale of livestock – were lacking and negative evidence is equally important.

The case of Leribe

An abbreviated description of the drought-relief programme in one district of Lesotho illustrates this point further.

The northern district of Leribe (see Fig. 1.1) was not originally intended to be on the GOL's list of seriously affected drought areas and was only subsequently targeted for relief due, in part, to reports of worsening conditions as the drought continued and also to administrative delays in other, worse affected districts, in completing acceptable beneficiary lists. Leribe could be said to have received food aid somewhat by default. There was no sound information on how many families had access to remittance money nor how much remittance money was available at any one time for food purchases. However, even a cursory look at food production, food aid receipts, and remittance money for the district as a whole would suggest that, even at the height of the drought, Leribe was perhaps more favoured than other districts, particularly when compared to adjacent mountain regions.

Table 1.3 shows domestic grain-production for three districts, including Leribe, for 1982/3 and by dividing these tonnages with the *de jure*

TABLE 1.3. Estimated grain harvest for three districts, 1982/3

	Production (tonnes)	Population	Kg/Head/Annum
Leribe	30 661	246 100	124·5
Butha Bethe	7 923	88 600	89·4
Quacha's Nek	1 504	88 300	17·0

Source: *Annual Statistical Bulletin* (1983). Government of Lesotho, Maseru, 1985.

population figures for the same period, an approximate estimate of kilograms per head can be arrived at. The results suggest that, of the three districts, Leribe was, very nearly, self-sufficient (i.e. 124·5 kg/capita from domestic grain-production alone, compared with the UN estimated requirement of 167 kg/capita/annum). Table 1.4 summarizes the ideal food grain requirement and matches this with the resources available in 1983. The food gap was estimated at 2357 tonnes which could well have been met through the production, consumption, or sale of vegetables and pulses, or the individual purchase of grains either locally or from the RSA.

It is difficult to provide a detailed breakdown of regular food aid directly allocated for Leribe. However, some estimates can be made using World Food Programme, Save the Children Fund and Catholic Relief Services figures. Table 1.5 lists the known quantities of food distributed through the above donors in 1983/4. The total regular food aid amounts to 2964·9 tonnes. The beneficiary lists indicate that 78 274 (31·8 per cent) of the *de*

TABLE 1.4. Estimated food requirements and food resources theoretically available, Leribe, 1983

		Tonnes
1. *Food Requirement** 246 100 × 167 kg	=	(41 098)
2. *Food Resources†*		
Domestic production (grains)	= 30 661	
Commercial imports	= 1 545	
Regular food aid	= 2 964·9	
	35 170·9	
3. *Food Gap*	=	(5 927·1)
Drought relief	= 3 570	
Balance	=	(2 357·1)

* This is the *de jure* rather than the *de facto* population figure. At any one time as many as 15 000 may be resident mine-workers in South Africa.

† Note that these figures do *not* include vegetable or pulse crops, nor any animal products.

TABLE 1.5. Food resources for Leribe District, 1983

Domestic production (grains) tonne	Commerical imports tonne	CRS regular programmes tonne	WFP/SCF regular programmes tonne	Total tonne	Kg/head
30 661	1545	1679	1285·9	35 170·9	142·9

Sources: *Annual Statistical Bulletin* (1983). Government of Lesotho, Maseru, 1985; CRS Call Forward Records (1983); SCF Statistical Returns (1983); *Food and Nutrition Information Bulletin*, No.1:3. FNCO.

jure population was in receipt of food supplements through school or institutional feeding programmes, food-for-work projects, or through clinics. It is worth noting that those registered on food-for-work projects receive rations for themselves *and* for up to five family members and that institutional feeding programmes provide three meals a day.

In addition to locally produced and imported food, the third resource available is cash. Leribe has the third largest per capita remittance income of Lesotho's ten districts. Although it cannot be assumed that cash is equally distributed or abundantly available in times of need, nevertheless the amount of cash theoretically available in a given area does give some indication of general economic activity and opportunity. If the remittances recorded for Leribe in 1983/4 are broken down the cash resources represent sufficient money to purchase 156·4 kg food grain per head per year (see Table 1.6). Grain prices were calculated on the basis of official prices and estimating an average for the three main crops of wheat, maize, and sorghum.

The government itself in 1984 reported that the greatest shortfalls were recorded in the mountains of the east and in the south, and that only Leribe showed a level of production close to the average.

Preliminary Conclusions

Lesotho does not have a centrally planned economy with a large dependence on markets in rural areas. Thus, a shortage of food due to crop

TABLE 1.6. Approximate grain-purchasing power: Leribe District, 1983

Average price of grain per kg (maize, wheat, sorghum)	Remittance per capita	Purchasing power (kg per head)
0·28/kg	43·8 maloti	156·4

Source: *Annual Statistical Bulletin* (1983). Government of Lesotho, 1985.

failures does not automatically render a large section of the population destitute. Local food shortages can be offset by individual purchases in the RSA. In a food-import economy famine would only occur if there were an equally sudden decline in imports or a drastic reduction in cash income. Neither of these events occurred between 1982 and 1983. Incidentally, had there, in fact, been famine the mortality rates would have been extremely high because emergency food aid only began distribution in some cases fully 18 months after the emergency had been declared. Preliminary conclusions suggest, therefore, that famine in a cash-based economy with regular food imports is unlikely. This does not, however, imply that individual households, especially, those without access to regular cash, cannot and do not starve. They do but this is not a definition of famine and requires a different aid response. Although one cannot assume a pervasive mechanism for the transfer of income between households, there are opportunities for individuals to earn money or food, and in this sense the Basotho are better-protected than many other communities in drought-prone Africa. The willingness of the donor community to respond to Lesotho's declared need was influenced by political considerations including the USA government's wish to prevent further compromise with the RSA and to limit the increasing influence of the Eastern bloc countries represented in Maseru.

Ideally, any aid input even in times of emergency should attempt to reduce rather than increase vulnerability. In Lesotho there exist many local coping-strategies, and although clearly compromises, nevertheless provide some protection to individual households. One may, therefore, legitimately inquire whether or not the kind of emergency-response in Lesotho helped to support these local strategies in, for instance, increasing transfer of income between households and providing added social security measures guaranteeing some degree of protection for the destitute.

Agricultural practice has its own mechanisms for distributing food and cash and any intervention which blocks these life-saving mechanisms is necessarily counter-productive. Although there is not as yet any direct evidence (pending future evaluation studies) it seems probable that the provision of free food may well have discouraged individual farmers from investing in seeds, fertilizer, and tractor hire for that season. A reduction in agricultural yield can in turn only limit the opportunity for the landless to acquire food through hiring, sharecropping, and work-party arrangements. Furthermore, a lack of surplus sorghum will severely cut back beer-brewing (and pig-breeding) which has been noted as an extremely important income source for women. The only advantage of reduced production is that fields may lie fallow for a season and thus have a small chance to recover a degree of fertility. It could also be argued that money saved on agricultural costs can accumulate for capital investment in the following season. But the real need in Lesotho, beyond the immediate, is for income generation within the country in order to prevent further dependence on the RSA and to limit the social disruption that migrancy causes. Free food aid does not fulfil either of these requirements nor does it necessarily benefit those who most need social insurance or security.

The major objection to the generous response to Lesotho's emergency was that it was possible because of the lack of information; the same lack of an information base resulted in food being distributed to those not necessarily in need and thus both the humanitarian as well as developmental potential of the response was forfeited. Finally it may be said that Lesotho has an extremely fragile economic base but because of its diversity and cash orientation is less immediately vulnerable than those communities which truly rely on subsistence agriculture in times of drought, or in which there has been less time to adapt to the social and economic effects of widespread migration.

MOZAMBIQUE, 1982–5

The Emergency

The drought which affected Lesotho similarly touched on several provinces of Mozambique shifting from the north to the west and then to the east during the period late 1981 to the end of 1984. However, the effects of the drought on Mozambique were profoundly different and this in turn was due to the peculiar historical factors which have shaped Mozambique's almost destitute economy and the increasing spread of guerrilla war. The Popular Republic of Mozambique, hereafter referred to as the Government of Mozambique (GOM), achieved independence in 1975 and immediately instituted a series of social and economic reforms aimed at radically altering both the means of production and the health and education services. The country was ill-prepared for independence in terms of available trained manpower, capital equipment, and access to large-scale development assistance. In the 10 years since independence, the GOM has had to contend with almost insurmountable economic problems, the growing impact of insurgent activity, drought, and cyclone.

Mozambique, although subject to periodic drought particularly in the south is a rich and, potentially, fertile country served by over 20 major rivers (see Fig. 1.2) and, in many areas, with land regularly capable of bearing two crops each season. Much of the surface area falls into the category of scrub land but the overall population is still sufficiently low to preclude high-density areas with extreme overburdening of the land, such as one sees in northern Ethiopia or Lesotho. There is no shortage of land and entitlement to land is gained simply by cultivating it. Mozambique

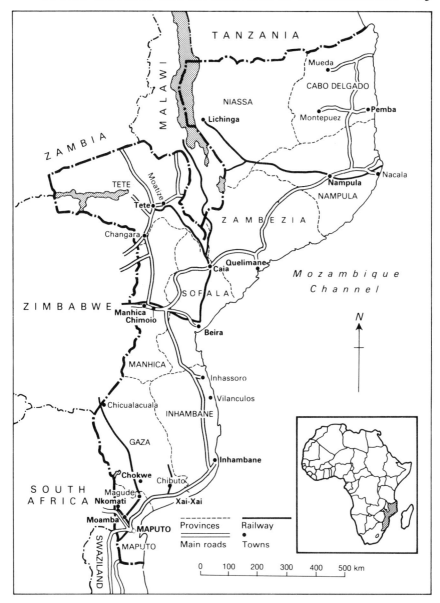

Fig. 1.2. Mozambique: provincial boundaries and communications.

also, uniquely, has over 12 important coastal ports, 3 of which are deep-sea harbours. Nevertheless, by far the greater majority of the rural population was, and remains, absolutely poor with an average annual per capita GNP of less than US$113, and an increasing dependence on aid and foreign

TABLE 1.7. Social and economic factors

Population	13·3 million
Growth rate	3·0%
Per capita GNP	US$128 (1982); US$125 (1983); US$113 (1984)
Infant mortality	240/1000
Average life expectancy	46 years
Literacy rate	10–20% (pre-Independence 3%)
Rural population	>80%
Number of migrant workers	40 000 (at Independence 120 000) (estimate)
Cereal imports	298 000 tonnes (1982)
Cereal aid	126 000 tonnes (1982)

Sources: FAO (1984), Maputo; Unicef (1985), Maputo.

inputs for survival (see Table 1.7). The negative growth of the economy as a whole during the last decade has many and complex causes but it has to be stressed that the resources available at the time of Frelimo's victory were extremely limited and international support in the following years never sufficient. Table 1.8 illustrates the prevalence of a subsistence economy prior to independence and Table 1.9, showing the exponential increase in food imports between 1974 and 1978, bears testimony to the

TABLE 1.8. Estimate of average land holdings pre-Independence

ha		% of the population
<0·5	=	20
0·5–1·0	=	40
1·0–4·0	=	35
>4·0	=	5

Source: Wuyts (1978).

TABLE 1.9. Food imports, pre-Independence, 1977–8

	pre-Independence (average)*	1977 tonnes	1978 tonnes
Wheat		96 700	121 400
Maize		52 000	99 300
Rice		37 300	84 200
TOTAL	110 000 tonnes	186 000	304 000

* Mozambique was never self-sufficient in wheat but the modern sector supplied 90% of rice consumption and over 50% of maize consumed by the urban population.

Source: FAO (1984), Rome.

exploitative Portuguese regime which had, over centuries, evolved a system of production including forced labour in the large agricultural and cotton farms of the north and which systematically excluded the peasants' participation in profits.

The drought which began in Zambezia and Nampula provinces in 1981 shifted to the south in 1982 and then affected the greater part of the south and centre of the country by 1984 and combined with the war, which intensified in 1984, devastating large areas of Mozambique. These two events, war and drought, together drastically reduced agricultural production and severely set back the GOM's plans for revitalizing the rural production especially in the family sector. An example was the cereal import requirement for 1985–6 which approached those of countries such as Ethiopia having nearly three times the population.

The Colonial Period

However, it is this chapter's hypothesis that factors such as war and drought are not of themselves sufficient conditions for the emergence of famine. Mozambique's vulnerability has its roots in a long period of colonialism which subjugated local and traditional agriculture in the interest of cash cropping and a lucrative export market. The massive exodus of Portuguese at independence left Mozambique neither able to continue an export-orientated agriculture system nor with the administrative infrastructure to implement immediately a more locally oriented agriculture. The constraint on local production initiatives by Portuguese planning forms a necessary background to understanding the present vulnerability.

Although the Portuguese presence in Mozambique dates back to the early seventeenth century, overt and widespread involvement in the country's agriculture only really began in the late nineteenth century when the potentially valuable markets in Europe for cotton and other raw materials were perceived. In addition a fast-growing urban population was increasingly serviced by settler farms in the surrounding rural hinterlands.

Broadly, there were three systems of agricultural production between the late nineteenth century and independence. The emphasis in the south (Gaza, Maputo, Manica, and Inhambane) was on the provision of labour for the mines in the RSA and in, what was then, Rhodesia; in contrast, there developed a plantation economy in the central region (Sofala, Tete, and Zambezia) and a peasant economy in the northern part of the country (Niassa, Nampula, and Cabo Delgado) (see Fig. 1.2).

The South

The effect of this colonial organization of Mozambique's agriculture was most visible in the disruption caused, primarily in the south, by the exodus of migrant labour. Although the system pre-dates Portuguese control, it was not until the late nineteenth century that substantial numbers of Mozambican peasants were employed in the RSA (12 000 from Inhambane alone in 1894, and an estimated total of 60 000 by 1900). In 1887 the British Consul reported that the natural produce of the area was nil, and wealth consisted of savings from migrant wages (see Ishemo 1978; Wield 1978).

Although some capital did accumulate, a reasonably well-off peasantry did not develop for a number of complex and interacting factors. These included the existence of a serious labour-shortage. This was due to the fact that fertile land was abundantly available, and that therefore, in the context of the major migration of labour, peasants with larger than average landholdings were unable to attract labouring assistance. Competitive wages were of no avail in a, by now, labour-based economy.

In addition, as in Lesotho, mining contracts of 12 months minimum or 18 months maximum were in no way geared to the agricultural cycle. As a result traditional cultivation, irrigation, and soil management practices fell into disuse (see Srivastara and Livingstone 1983). Although a small peasant producer-class did eventually emerge, by far the greater number of households in the south were unable to produce sufficient to sustain livelihood and thus remained heavily reliant on cash remittances. Owing to the labour shortage, opening up new land in frontier areas was only a possibility to those who had tractors and ploughs, which were extremely few. Cash cropping among peasants was not fully developed until the 1960s when, with the end of forced cultivation, there came an expansion of coconut ownership and cashew cultivation and processing.

The Central and Northern Regions

Meanwhile in central and northern regions the economy was based primarily on the operations of foreign companies and individuals whose large estates were cultivated by a labour force ensured through the *chibalo* system – in which the government enforced or indented labour. Of 2000 large farms, 400 were owned by foreign commercial enterprises, concentrating on cash crops such as cotton, copra, and sugar. In addition, peasants were themselves forced to grow certain crops which were then bought at grossly exploitative rates.

Commercial Relations

Although the peasantry in Mozambique continued throughout the period up to independence to be the principal producers, little, if any, of the accrued profits benefited the subsistence farmer. The economy was almost wholly export oriented. The peasant's capacity to increase production was severely limited due to the factors already outlined above: low wages, forced labour and cultivation, and enforced migration leading to a labour shortage. And although land was available, an almost pathetic lack of implements curtailed opening-up of much new land. Unlike Lesotho, the total area of land under cultivation, even today, remains small in relation to land available. A decline in subsistence farming yields due to the above factors, in turn, encouraged increased migration in search of a cash income.

Nor was there the opportunity to develop indigenous marketing systems due to the presence of *cantinheiros* or merchant traders who, for at least a century, had monopolized internal trade. *Cantinheiros,* of both Indian and Portuguese origins, assiduously maintained beneficial terms of trade for themselves and, through their connections with banks and their money-lending functions, 'ultimately also had a share in appropriation of peasants' surpluses' (Srivastara and Livingstone 1983, p. 253). For example, Ishemo (1978) reports that the *cantinheiros* in Zambezia and Tete provinces developed the practice of paying for oil-seed crops before they were harvested, thereby ensuring the lowest buying-price.

Until World War II, the Portuguese Government's income was largely derived from mining remittances, revenue from the railway trade, and profits from land leasing transactions. After the war and up to independence, there was a greater effort to become more directly involved in cash crop production but none of these activities served to develop the rural peasantry or to create administrative, marketing, or communications networks. Social investment in the rural sector was negligible. For example, it was reported that by the early 1960s there were only 2800 rudimentary schools, almost all run by Roman Catholic missions. Endemic malnutrition was prevalent on a basic diet of maize and cassava.

To sum up, Mozambique's task at independence was therefore a formidable one. To add to the enormous economic burden and the ravages of a 10-year liberation war, several other events coincided with independence to create a virtual emergency in the first years.

Firstly, there was an astonishing flight of capital and skills in the year preceding independence leaving, some say, only 30 qualified doctors in the country. Secondly, South Africa, unable to get out of its commitment to the new government to pay for labour, in part, with gold, reduced the number of migrant workers from 120 000 in 1975 to 40 000 in 1976. Thirdly, revenue from the railway trade was also substantially reduced with South Africa preferring other outlets. A fourth factor was that plantation work in what was then southern Rhodesia also ceased and there was massive migration of peasants from rural to urban areas in search of work. To add to this, Mozambique suffered several severe floods (Lim-popo, 1975 and 1977; Incomati, 1976; Zambezi, 1978) in the early days of independence, resulting in large displacement of local populations and destruction of crops. Finally, in 1980 when Zimbabwe became independent, the army created by former Rhodesian Prime Minister Smith to fight black liberation armies gathered strength with the help of foreign funding, and began a strategy of destabilization in Mozambique. The war waged by the Mozambique National Resistance (MNR) has been and remains the most severe threat to food security.

The Current Context

Agricultural Organization

Following independence, agricultural production was organized into four main sectors each having an annual production target, and some sectors receiving more state support than is the case today.

A total of 2·2 million ha is thought to be under cultivation of which 140 000 ha (6 per cent) was allocated to state farms, employing an estimated 150 000 workers. The state farms accounted for more than 90 per cent of the GOM's investment in agriculture and, together with private farms, are the only sectors where modern agricultural techniques such as a degree of mechanization, fertilizers, and pesticides are used. The private farms occupy approximately 50 000 ha and grow predominantly cash crops. This sector has clearly become less significant in terms of the amount of land cultivated and in production figures. Co-operative farms remain a limited sector in spite of government hopes and support and, at present, are thought to cultivate only 19 000 ha. The family sector of farms is by far the most important group covering about 85 per cent of the total cultivated land and with perhaps 80 per cent of the rural population entirely dependent on its produce. All cassava is grown within this sector as well as the bulk of cereal and cashew crops.

During the past 10 years the GOM has attempted to transform the family sector into communal villages (aldeias comunais) and in 1984 it was estimated that approximately 2 million people were thus organized. Since the intensification of the war and changing leadership at village levels, many of those newly liberated from the MNR and those returnees from Zimbabwe (where villagers attempted to escape drought), are organized in communal villages. The production figures for this sector are not known, but the few surveys that have been carried out reveal poor nutritional conditions, at times requiring air lifting of emergency food.

This framework for agricultural production was followed resolutely until 1983 when the fourth Frelimo Party Congress announced its intention to give greater priority than before to the family sector. This was primarily due to the fact that production within other sectors had consistently fallen well below even modest expectations. Thus the drive for socialization of the rural population along centrally organized five-year plans gave way to a degree of decentralization including the encouragement of small community-based projects, extra support for the family sector, lesser emphasis on state farm development, and more authority to provincial and district governments. However, efforts to expand communal villages continue.

The state marketing system is run by Agricom. This is a commercial enterprise which operates at rural level through its agents to purchase surpluses and to supply farmers with essential items of industrial origin. Agricom has its roots in the *cantinheiros* system and although the GOM's intention was to set up a nationwide marketing network, these plans have been greatly thwarted by limited access to rural areas because of the war and the extreme shortage of consumer goods. Agricom itself estimates that, at present, it can barely supply 10 per cent of the need. Incentive for farmers to invest in seeds, and possibly fertilizer, for an increased yield in the absence of any consumer goods is, thus, low. Additionally the flourishing black market in food is such that those farmers who are able to produce a surplus clearly would prefer to sell to private traders rather than to the State which has both fixed and low prices. The official buying price of maize is set at 13 metecais per kilogram whereas on the black market (*candonga*) a farmer can get 90 metecais for a kilogram of maize.

The Evidence for Famine

GOM Estimates

It should be emphasized that the government machinery for collecting and analysing information was, and is, severely hampered by lack of access to local areas because of insurgency, by a lack of expertise at central level in interpreting information, and by a breakdown (in some areas, complete) in communications for either conveying emergency information or for delivering material relief. In spite of these considerable handicaps, however, a picture of extreme distress did begin to emerge.

In early 1983 the GOM estimated that up to 4 million people were at risk in Southern and Central provinces due to drought and insecurity. In June 1983 a second report was issued on the worsening situation and the prospects for the next harvest. In fact, the June 1983 harvest was an almost total failure in the Southern provinces and yields were severely reduced in Central areas of the country. The GOM feared that the 1984 harvest would be equally catastrophic due to lack of rain and seeds.

By July 1983 the IDC was sufficiently alerted to plan assessment missions, although at least one mission subsequently reported that the situation in Maputo province was 'grave but not critical'. It was generally known that food was scarce in Maputo, Gaza, and, especially, in Inhambane. In September 1983, the International Committee of the Red Cross conducted a nutritional survey in Villanculos and recorded grave malnutrition. In October 1983 the Christian Council for Mozambique reported deaths from starvation in the provinces of Gaza and Inhambane and simultaneously pictures of dying children began to appear in the national press.

The Center for Disease Control (CDC, Georgia, USA) carried out a survey of drought-affected areas in October 1983 and the report was published in November of that year. The report stated that 6 per cent of children measured were less than 70 per cent wt/ht; that 22 per cent of these were below one year of age and 12·5 per cent of the 1–5 year age-group were dying in relief camps in and around Villanculos. The situation was compared to the worst stages in Biafran, Sahelian, and Somali crises of earlier years. In the same month a visit by Ministry of Health officials, Unicef, and the Mozambique Red Cross to Gaza concluded that there were high levels of malnutrition in accessible populations.

Calculation of numbers affected suffered all the difficulties that one might expect. Apart from some towns in the coastal areas large parts of the country were virtually cut off due to insurgency. Coupled with this, the lack of air transport and fuel made it extremely difficult for agency staff to travel. The GOM's January 1984 report indicated that 6 out of 10 provinces were affected and that 39 per cent of the total population was at risk of which 12 per cent were at severe risk of starvation. These figures were based on a slender information basis largely gained from local government officials and relief agency surveys from the affected areas. Given that insurgency precluded surveys outside the towns, extrapolations had to be made. These were based on the 1980 census data which included figures for rural and urban populations in all the provinces. Observers at the time reported that people were travelling as much as 100 km from the interior to reach the coastal food-distribution centres. There was also a belief expressed by relief agencies that those arriving at the coastal feeding centres may have represented the first wave of famine refugees and were, thus, the tip of the iceberg.

Survey Results

The purpose in presenting the following surveys (see Table 1.10) is to use the results to construct a picture of vulnerability in Mozambique. Although many different kinds of surveys have been carried out in the years following the declaration of an emergency in early 1983 by the GOM, the majority of them either do not record social and economic data or responses to drought and food shortages, and the few that do show no real understanding of the necessity to note *unusual* responses. The following analysis only uses the results of surveys which were more comprehensive in both methods and the scope of the enquiry.

Social and Economic Vulnerability Four separate surveys were undertaken in Changara: (Ntemangau) Tete, March 1985; Espungabera,

TABLE 1.10

Province	District	Date	% Malnutrition		Drought in previous six months
			wt/ht	ht/age	
Tete	Misaua	April 1984	12·0	–	dry
	Ntemangau	March 1985	5·3	29·8	dry–normal
	Mutarara	May 1985	1·2	–	dry
	Estima	December 1985	1·0	–	normal
Gaza	Chibuto + Manjacaze	October 1983	2·8	–	dry
	Chicualacuala	January 1986	4·4	16·0	N/K
Inhambane	Vilanculos	October 1983	7·8	–	very dry
Manica	Espungabera	March 1985	2·8	–	very dry
Zambezia	Quelimane	July 1985	5·9	18·1	normal

Sources: Unicef (1985); GOM, Ministry of Health, Nutrition Section (1985); GOM, Meteorological Service (1983–6).

Manica, March 1985; Quelimane, Zambezia, July 1985; and Estima, Tete, December 1985 (see Fig. 1.2).

In Changara (Ntemangau) and Espungabera (Unicef) the surveys were carried out in order to plan integrated rural development schemes and concentrated on identifying the poor and vulnerable as well as the resources available in the community in terms of skills and infrastructure. Changara was selected as an area for further input precisely because local initiatives to combat drought in its earliest stages had been impressive but had ultimately failed for want of material inputs to sustain community water and food management schemes. The Espungabera programme was directed towards the development of communities newly arrived in the area either from Zimbabwe where they had fled to escape drought and insecurity or from nearby zones recently liberated from the MNR. This survey was therefore dealing with so-called 'vintages' of displaced people, some of whom had been settled for more than a year and some who had arrived and continued to arrive in the weeks immediately preceding survey work. This constant influx of people, although a methodological nightmare, provided valuable data on the effects of displacement on nutritional and general health status of people.

Both the Quelimane (GOM, Ministry of Health) and Estima studies were primarily directed at measuring nutritional status of children but in both surveys serious attempts to understand the causes of factors associ-ated with severe malnutrition, were made. These studies are particularly valuable in that one of them presents a picture of *urban* vulnerability

(Quelimane) and both attempt to define high-risk groups in terms of assets or the lack of them.

A preliminary list of associated factors which categorize high-risk groups can be compiled. The results of the Changara Survey distinguish the characteristics of those households which were affected at the beginning of the drought and also of those households which were more severely affected. In each village there was a range of very poor to relatively well-off households and this clearly had an effect on the nutritional status of children. The retrospective analysis indicated that the drought had the most severe effect, resulting in deaths, on those households composed of predominantly old people, i.e. those who had no productive family members, and those households which had no access to low-lying land or cegonhas.* Poverty in the post-drought phase was characterized by small and ill-maintained houses, families with members working abroad (Zimbabwe) who did not send sufficient food or money, and absence of a productive work force. These factors were associated with poor nutritional status of children.

Conversely, relatively good nutritional status of children was associated with a degree of wealth characterized by a large number of family members who were wage-earners, and households in which children had completed vaccinations. This in turn was linked to the proximity of a health centre or post and to *machambas* (farms) not too distant from houses thereby allowing women time to visit health centres regularly. It is also suggested that families which attend are more settled and stable.

The results suggest that drought can in most communities be survived, up to a point, within the context of a reasonably stable social network; people with land near to the river or with cegonhas were prepared to allow others access to water and equipment. If, however, these local efforts become overburdened due to lack of materials to divert water, or because social networks are suddenly disrupted – people can begin to starve with astonishing speed.

Protection from starvation is associated with having a range of options (in the Changara study, animals, cash income, land, remittances, agricultural implements, sufficient manpower, etc.) which in turn is related to having a more diverse source of income. Poverty arises from dependence on a single and uncertain source of income and no means to extend the range of choices. Diversifying one's source of income in the Changara and other contexts means varied seasonal production and this can only be achieved with a relative degree of physical and social stability. Therefore the priority demonstrated in this study is to deal with those factors which cause instability such as migration.

---

* Local method of irrigation, literally a 'stork'.

This theme is further illustrated in the Espungabera survey results. All children below 20 kg weight in four villages were weighed and measured. Nutritional status in children was linked with the household's access to quantity and quality of food, and disease. The population under scrutiny was composed of peasants who had either returned from Zimbabwe in the past year or had been recently liberated from MNR-controlled areas. It was noticeable in the survey results that those villages most recently established and, thus, having the greatest number of socially dislocated people, also had the greatest prevalence of malnutrition and disease. The result again emphasizes the crucial role of social dislocation as a cause of extreme stress and starvation.

The Estima survey was able to distinguish the vulnerable from the less vulnerable on the basis of number and kind of animals per household. The poorest had either no animals at all or less than five pigs or goats and some chickens and ducks. At the other end of the scale, the wealthy (i.e. those with assets to trade in time of need) owned more than 10 pigs or goats or larger animals, particularly plough oxen. On the basis of this kind of information a clear difference between villages was visible and the poorest had the highest rate of malnutrition, the lowest number of wage-earners and the least amount of household food stocks.

In Quelimane the survey included in-depth family and clinical histories. Severe malnutrition requiring hospitalization was associated with families in which the husband had more than one wife, a low salary, large number of children, small *machambas*, and a lack of variety in crops produced.

Relationships Between Displacement and Starvation The suggestion that insecurity and displacement are more serious causes of vulnerability and even starvation can be further tested by attempting to put together the results of surveys which show alarmingly high rates of malnutrition with the distribution of (a) drought and (b) war. This necessarily has to be a limited exercise because nutritional surveys, for obvious reasons, are largely undertaken in relatively secure areas, rainfall distribution charts are not uniformly distributed or always reliable, and areas of insurgent activity vary from month to month as well as in the severity of the disruption that they cause. Another important limitation is that *credible* nutritional surveys are few and far between.

Some explanation of methods used is called for. Anthropometric or body measurements distinguish between current (wt/ht) and past (ht/age) undernutrition. Nutritionists broadly agree on cut-off points for each type of measurement, that is reference figures below which children are considered to be moderately or severely malnourished.

Table 1.10 lists the malnutrition rates (only severe malnutrition has been reproduced) found during the surveys and combines this with general evidence of drought in the previous 6-month period. Taking the nutritional results first: evidence of the most severe current malnutrition (i.e. wt/ht) comes from:

- Vilanculos, Inhambane in October 1983
- Misaua, Tete in April 1984
- Ntemangau, Tete in March 1985
- Quelimane, Zambezia in July 1985

The highest rates of past or historical malnutrition (i.e. ht/age) are also seen in Ntemangau. The least-affected areas appear to be Estima and Mutarara, in Tete Province.

Meteorological records (see Table 1.11) clearly show that the worst affected province between 1980–3 was Gaza which has had only one relatively normal rainfall year (1981) between 1980–5. Inhambane was badly affected by drought in 1983 as was Maputo Province. However, using these official records, Tete was one of the *least*-affected provinces. This, in so far as it is permissible to manipulate the data, is an interesting result. It has already been shown that Tete Province as a whole has both the worst and the least incidence of malnutrition in children; it may, therefore, be justifiable to look for causes other than drought.

Fig. 1.3 shows areas of insurgent activity and concentration just west of the Tete border with Malawi. In particular this map shows evidence of gross disruption to communications which is a well-documented cause of sudden and severe food shortages in highly localized areas. This is particularly relevant in northern Tete where, traditionally, the fertile Angonia region has supplied other more southerly areas with grain in return for manufactured and consumer goods. This trade, well established in the region, allowed a degree of specialization in each area which ultimately made people vulnerable when communications were so abruptly disrupted and trade was either prohibitively expensive or no longer possible, because of insecurity.

Insurgent activity, in that it provokes populations to move to wherever they can find relative safety, is clearly a serious and ongoing cause of hunger and distress. Even as early as 1983 before Inhambane began to suffer the cumulative effects of drought, the poor nutritional conditions recorded in October of that year may well have been partly caused, and certainly exacerbated, by war in neighbouring Sofala and northern Inhambane which created a flow of shocked and impoverished people to the coastal areas.

Gaza has suffered the most intense drought for the greatest number of years and the cumulative effect is evident. The malnutrition recorded in recent surveys is highest in those most recently arrived in the resettlement camps. However, as can be seen from Fig. 1.3 war was well established in

TABLE 1.11. Mozambique: summary of drought conditions

Province	1980								1981								1982								1983							
	O	N	D	J	F	M	A	M	O	N	D	J	F	M	A	M	O	N	D	J	F	M	A	M	O	N	D	J	F	M	A	M
Maputo	×	×|	○|		×		×													×|	×|	×|	×|		×	○	○	×|	×|	×|	×|	
Gaza	×	×|	○|	○|	×	○	×													×|	×|	○|				○	○	×|	×|	×|	×|	×
Inhambane	×	×	○|	×			○													×|	×|	×|				○|	○|		×	×|	×|	
Manica		×|		×	○|																	×|					×|	×|	×			
Sofala		×|	×	×	×	×	×	×										×|							×	×	×			×		
Tete	×	×	×	×	×	×				×	○|	○						×|	×|								×|	×	×		×	
Zambezia		×	×	×	×	×	○|	○|		○|	○	○	○					×|	×|								×			○	×|	
Nampula	○	×	○|	○	○	○	○				○	○						×	×				×	×							○	×
C. Delgado	×	×			○|	○	○	×			○	○	×|																○		○	
Niassa		×|	×	×	×	×	×			×	×|									×|		×					×	×	×			

× Dry in part of the Province
○ Very dry in part of the Province
×| Dry in the whole Province or most of the Province
○| Very dry in the whole Province or most of the Province

Source: Meteorological Service of Mozambique, 1980–3.

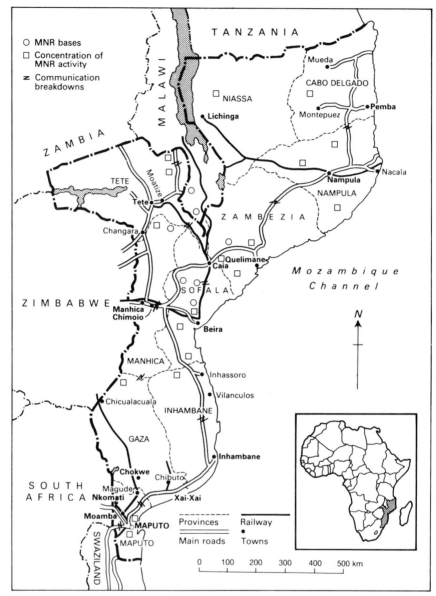

Fig. 1.3. Mozambique: insurgent activity 1983–4.

the region since at least 1983 and it is possible that it was the combined effects which drove people to seek refuge in Zimbabwe. Given the prevalence of war it is not possible to know without more evidence if drought alone forced people to move.

The high rates of current malnutrition* in Quelimane are interesting given that this part of Zambezia has apparently not suffered severe drought since 1980. However, regions surrounding Quelimane have seen much and increasing amounts of armed fighting since 1983 with consequent disruption to communications. The results would suggest that food shortages and other deprivations can be just as severe in urban as in rural areas and that virtual siege economies affect the urban poor seriously.

In many economically depressed areas of Mozambique, successive seasons of drought may well have triggered a crisis, but the war is and has been a most important contributory factor.

These are not startling conclusions; they merely illustrate that there is a syndrome of poverty and vulnerability which can be identified. If properly identified then the allocation of scarce resources can be a more logical process. Further than this, however, it does seem important to acknowledge that drought does not necessarily provoke famine. Repeated crop failures have clearly made people vulnerable but centuries of drought-occurrences have moulded local coping-responses. When these responses become too overburdened, people migrate in search of jobs and food. It is *this* response which tends to make people (migrants as well as those who remain behind) highly vulnerable in the short, as well as in the long, term. Therefore famine prevention, and rehabilitation programmes, must, as a priority, *prevent* migration and this requires sufficiently early and, perhaps only limited, action to support existing coping-responses. Food aid may be needed to tide some families over to the next harvest but the greater majority would benefit more from material inputs to sustain irrigation of crops or other coping-responses.

Preliminary Conclusions

Mozambique is a fertile country with no lack of land or, in times of peace, restriction on increasing landholdings. The constraints to increased or even adequate production in Mozambique at present are chiefly concerned with lack of capital resources with which to exploit land, lack of any kind of sufficient marketing structure to ensure a balance of food between surplus and deficit areas or to encourage farmers to produce surpluses; this includes the chronic and severe lack of consumer goods with which to trade. But perhaps the most severe obstacle to subsistence farming, at present, is the widespread insecurity which has caused and continues to cause massive dislocation. The survey results given above show without a doubt that those who are uprooted rapidly fall into destitution and thus it

---

* It should be said that evidence of malnutrition in children is not necessarily due to food shortage, but can also be caused by a variety of other factors including disease, ignorance, and neglect.

could also be argued that the drought, although it undoubtedly exacerbates extremely vulnerable communities, was not the cause of famine.

It has been said earlier in this chapter that war is not necessarily a precipitator of famine. The case of Cambodia was quoted; there are other examples where resilient farmers can for at least a period manage to continue planting and harvesting. However the peculiar conditions of war in Mozambique were and are particularly destructive to subsistence farming. The guerrilla war has no recognizable frontline and the insecurity and uncertainty that this provokes prevent already impoverished farmers from further investment in the land. The toll that this war takes on individual families' time and energy is considerable.

For example, in 1986 a previously safe area of Inhambane was suddenly attacked with heavy loss of life and savage injury. Villagers in the area fearing further night attacks, left their fields and households at four in the afternoon carrying as many movable assets as they could, such as agricultural tools, and trekked into the administrative centre where they camped overnight, leaving the next morning to return to agricultural tasks. The mileage involved in these double journeys could amount to 30 km per day (personal communication, Dr Julia Stuckley).

The costs of importing even essential foods such as salt, soap, sugar, and clothing are so high that the terms of trade each month become more and more to the disadvantage of the farmer. An informal survey (D'Souza 1986) of current prices in Inhambane indicated that the daily casual labour rate of between 100 to 150 metecais per day would buy 1 kg potatoes or a strip of tobacco or 3·5 kg of cassava. In spite of this cash is decreasing in value and rather a barter or exchange trade has developed which is extremely sensitive to any changes in supplies. Whereas in 1980 a cow could fetch between 20 000 and 30 000 metecais, in 1986 it was worth 35 kg of maize. In early 1983 a chicken could be sold in order to purchase enough maize for a family for a week whereas in 1985 1 chicken was worth 3 bars of soap.

Once again one can see that the family which has managed to retain assets and thus has the basis for exchange is more likely to survive than the family with no assets. A definition of poverty is the family which is small, female-headed, and without animals or farm tools. Again in the informal survey previously referred to, out of 65 household representatives only 24 had machetes (37 per cent), the basic agricultural implement, and of those without machetes, i.e. 41 households, 23 were female-headed. Those households which still have a foreign cash income from South Africa or Zimbabwe are relatively well off since the official exchange rate for the rand is 16 metecais per rand whereas on the *candonga* over 300 metecais can be obtained.

Mozambique suffers extraordinary poverty by any standards; there is not only widespread lack of entitlement to food and other essential goods but also a severe and absolute lack of food in certain areas caused by lowered production, poor transport, and a moving rural population. The fact that starvation is not more widespread than it is suggests that extensive networks for the exchange and sharing of food exist and also that the country itself is sufficiently fertile to allow people to live off cassava and wild foods. The aid requirement in Mozambique is no different from that in other vulnerable countries – namely the need to increase local production through the use of incentives (i.e. consumer goods and realistic pricing policies) and a cessation of the war to allow families to invest in their land. Free food will not necessarily achieve these economic and political imperatives. Relocation schemes, although sometimes a necessity, do not have a good record, in terms of agricultural results, in Mozambique.

The next case study, that of Ethiopia, will summarize the indicators of the famines which occurred in 1972–4 and 1983–4 and examines the potential of government schemes to reduce vulnerability.

ETHIOPIA, 1972–4: 1983–4

Background

The pre-famine indicators in Ethiopia are well documented and, sadly, these indicators were all too obvious in 1983–4. This section, therefore, will concentrate less on understanding the nature of vulnerability amongst Ethiopian peasants other than recording what was known and what was done with that information, but look more at government policy to reduce vulnerability, namely the relocation programme. It has been suggested in the previous two case studies of Lesotho and Mozambique that social disruption whether due to migrant labour or war, is economically destabilizing and can contribute to increased vulnerability. Thus it would seem legitimate to examine the potential of massive resettlement programmes, which though intended to give greater security to peasants, may in fact create a population far less resilient in the event of future drought. It might also be useful at this juncture to reiterate observations in both Lesotho and Mozambique that those households with an extensive kinship network and with diverse sources of income can withstand food shortages better than those in which there is reliance on a single source of income whether that be cash or food.

During the period 1972–4 there were two separate famines in Ethiopia, and for analytical purposes it is important to distinguish them. The first, at its height in 1973, affected the north and predominantly the Province of Wollo (see Fig. 1.4); the second affected the large south-east district of the Province of Harerghe and the impact of the 1973 drought was magnified because of structural changes affecting particularly the freedom of seasonal movement and grazing land of the Ogaden and Issa Somali herdsmen.

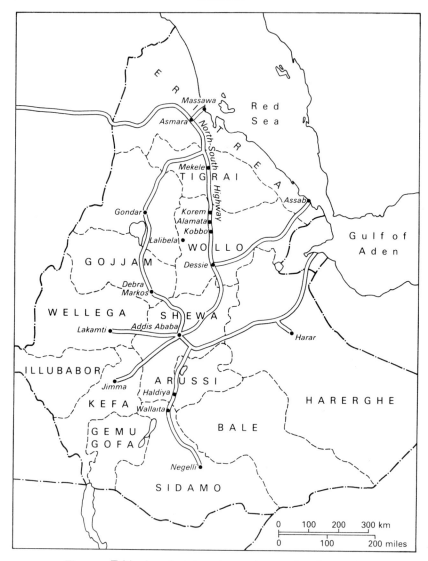

Fig. 1.4. Ethiopia: administrative regions and major roads.

Pre-famine Indicators in the Ethiopian Famines of 1972–4: 1983–4

1972–4

The failure of the main rains in mid-1972 followed by almost complete failure of the spring (Belg) rains in early 1973 came after almost 7 years of inadequate rain in parts of northern Ethiopia. Peasants already stretched to the limit of their reserves, succumbed to famine and starvation. Sen (1981) has argued persuasively that the famine was a result of entitlement failures because although there was undoubtedly a reduction of grain output in 1972, famine and starvation were precipitated by loss of wages in cash or kind. Sen publishes data collected at the time by the Ethiopian ministry of agriculture which suggest that of the 14 districts only 7 per cent had a substantially below normal output of food grains. He concludes that there was 'very little evidence of a dramatic decline of food availability in Ethiopia coinciding with the famine' (p. 92). Moreover, there is evidence that food from the worst-affected districts of Wollo and Tigrai was exported, although in small measure. This, Sen believes, represents negative evidence for those who would argue that famine was indeed due to a decline in food availability because of the lack of roads and transport to import food from surplus areas. The fact that this did not happen was, Sen says, not due to insurmountable logistic problems but rather because markets in Wollo had no purchasing power. This, in turn, was due to the fact that peasants had insufficient cash to create an overt market demand such that traders would judge it profitable to respond.

The categories most affected included pastoralists, evicted farm servants and dependants of farmers, tenant cultivators, small land-owning cultivators, women in service occupations, weavers and other craftsmen, and occupational beggars. In these groups an agricultural failure was equivalent to a collapse in income and thus of entitlement to food. The sudden breakdown in longstanding patron–client and servant relationships, whereby the richer landowners could no longer afford to feed such an extended household, coincided with a rapid decline in demand for services and goods provided by craftsmen, beer sellers, and the like, resulting in a large number of people with no assets and no entitlement. Livestock prices, in spite of the reduction of stock due to drought and poor grazing, fell further reducing the peasants', and most especially the nomads', purchasing power. Interestingly, however, market prices of food did not rise substantially and this, Sen argues, was predominantly due to the fact that entitlement to food in these communities was direct rather than by market transfer. Thus, although there was some increased demand on the market by those who still had cash, by far the greater number of peasants in the area had obtained food directly in lieu of agricultural or other services and thus remained outside the market sphere.

An analysis of origin and occupation of those in relief camps clearly indicated that those who came from densely populated areas and who were dependent on exchange of services for grain suffered worst. In Harerghe the picture was somewhat different. Substantial tracts of particularly valuable grazing land had been given over to commercial agriculture and this continued throughout the early 1970s. Loss of livestock due to displacement from traditional grazing land and the drought was compounded by a rapid and real decline in the exchange value of livestock for grain, upon which the nomadic pastoralists depend for about half their food requirement (see Seaman *et al.* 1974, 1978). Sen states that the percentage loss of grain entitlement due to the above factors was an astonishing 92 per cent for pastoralists in the Issa desert.

Neither the research community nor the government of Ethiopia or aid workers were attuned, at that time, to the significance of such fluctuations in people's entitlement to food and its collapse. Nevertheless, even with overt signs of famine and starvation, including the unmistakable crowding of starving people on the main north–south road through Wollo and mass migration to urban and relief centres, the relief community was slow to respond. Aid in adequate amounts only began to arrive in Wollo in 1974 by which time the famine was subsiding due to good mid-year rains in 1973. Meanwhile the Harerghe famine was at its height but bulk food-delivery from reserves in the north to the south-east proved very difficult and, thus, a second famine was not prevented, though, to some extent, it was alleviated.

1983–4

In 1983/4 the indicators were depressingly similar. Insufficient data on grain and livestock prices exist and will have to await, as was the case of the earlier famine, for retrospective studies. Nevertheless, the overt signs of impending famine rather than just increasingly dangerous vulnerability were visible as early as spring of 1984 whereas a response in terms of adequate food-commitment did not begin until the autumn of that year by which time Ethiopia was once again in the grip of most severe starvation (see Holt and Cutler 1984; Cutler 1984). Perhaps the salient difference between the previous famine of 1972/4 and that of 1983/4 was that due to more equitable land distribution and, thus, a reduction in the number of landless peasants, there was greater overall involvement in the market and thus the direct food entitlement failure resulting in reasonably stable food prices, changed. Food prices rose rapidly in drought affected areas. This gave rise to a number of responses such as increases in petty trading, influx of casual labourers on to the market and a concomitant depression in daily wage labour rates, great increases in the volume of livestock sales and, eventually, migration to urban and relief centres.

In addition, in the decade or so between the two major famines there were a large number of local emergencies precipitated by the continuing deterioration of land and agricultural production, as well as rainfall failures. The economic strategy of the government for rural areas, particularly pricing of surplus produce, has not encouraged farmers to increase agricultural production. This together with the war in the north and its consequent effect on inter-regional trade has seriously impaired the capacity of farmers in drought-prone areas to build sufficient reserves to withstand cumulative years of drought.

For example, during the 1973/4 drought a farmer from Wollo was able to take his cattle to the neighbouring fertile Province of Gojjam and exchange one ox for at least 500 kg of teff. However, current government restrictions on movements of cattle and food and also on travel meant that in the 1984 crisis one ox could be exchanged for only 20 kg of food grains. Small local food markets which continued to operate were confined to getting supplies from the immediate neighbourhood and local shortages caused huge price rises and also encouraged hoarding. Farmers became so desperate to raise cash that it was reported that some sold their houses as firewood in order to raise money to buy food. The fact of famine and starvation has had profoundly disruptive effects on communities not only due to very high mortality but the sheer impact of migration itself.

There is no doubt that once migration has begun as a result of local food shortages people become increasingly vulnerable to death by starvation due to a combination of physiological and psychological weakness as well as the increased incidence of communicable disease. And therefore any attempt to preclude distress migration is to be taken very seriously indeed. The following effects of migration, which in themselves render a community more vulnerable to famine in the future, have been recorded.

On the more practical side, because distress migration makes people so weak, there is a corresponding need for any relief measures to be extraordinarily efficient if large numbers of deaths are to be avoided. However, the evidence would suggest that this is very difficult to achieve because of logistic and other reasons and therefore there is an almost inevitable association between migration and widespread deaths.

Secondly, people who leave their home villages, and thus their land, in order to congregate around relief food-centres, which are often at considerable distances from their villages, are not able to take advantage of any sudden or early rains for planting and, moreover, if and when they do return to their land it will require that much more investment in order to make it productive once again.

A more long-term view would suggest that social units, though enduring and stable, have their limits. There are as yet insufficient anthropological studies of what are the necessary conditions for such social units to survive even under the greatest stress. Nor are there sufficient studies on social recovery on what might be the tensions and constraints in communities which attempt to come together again following trauma such as starvation, death, and dispersion. Inevitably, that society will be depleted in its members as well as in its material reserves. Depletion occurs not only because of the mortality due to starvation but also because younger members of that society may have been lured by the promise or even the rumour of employment elsewhere and therefore choose to abandon their farms (and even their families), either temporarily or more permanently in the hope of a less vulnerable and poverty-stricken way of life.

The Resettlement Programme

The indicators of increasing distress in early 1984 although noted by several field agencies were neither systematically recorded nor presented as a body of credible evidence to the IDC. In addition the Government of Ethiopia for vested political reasons partly to do with the Tenth Anniversary Celebration of the Revolution in September 1984 was unwilling to declare an emergency and certainly reluctant to allow access by outsiders to the affected communities of the north (see Cutler and Stephenson 1984). The famine is over and no doubt specialized studies to define the local pre-famine context will be undertaken and added to the general body of knowledge as to how and why famines occur. The immediate interest here, however, is to assess the potential for recovery and, in particular, to examine the Government of Ethiopia's plans for resettling up to 1·5 million people (later scaled down to 0·9 million) from the northern heavily-populated and infertile regions to more southerly, and potentially more fertile, regions in an attempt to reduce the burden of people on over-exploited soil. *Regardless* of the political aspects, of which there is no doubt there are many, the purpose is to understand if this kind of intervention might ultimately provide better protection against drought.

As at April 1985 the government estimated that up to 320 000 people had been resettled mainly from the northern provinces. The basic principles governing the resettlement programme include provision of a site, services, and sufficient food to the newly-arrived groups to bridge the gap between the preparation of the land, planting, and harvest. Additionally, the government believed that food subsidies to the newly arrived were necessary to prevent local market shortages and price fluctuations. The choice of resettlement sites was based on the selection of a surplus-producing area and where the local peasant associations were sufficiently robust to absorb new groups. The government made no formal distinction between the emergency and the rehabilitation stages of an operation and fully expected to use emergency food aid in speeding up resettlement of people from food deficit areas in the north to sites in the west and south-west.

Previous resettlement programmes in Ethiopia have been hampered by a number of problems including the following: selection of sites was not primarily based on agricultural, social, or health factors but due to political considerations; there was inadequate attention to the involvement of people themselves in making decisions about how communities should be organized in the new areas; and the technical and administrative inputs were overwhelming and tended, at times, to create a dependency attitude. The current resettlement approach claims to have recognized these constraints and to have incorporated solutions into future planning. However, field visits by some agency representatives would suggest that this may not necessarily be the case. Newly resettled people have been organized into village units too large (more than 1000 families in a given area) to allow individual farming, and this does not encourage integration with the local population but results in a clear separation of farming and grazing areas to the disadvantage of the settlers. Soils are too heavy to be cultivated by animal ploughing and, in addition, the soil has water-logging problems and requires fertilizer to be productive.

Social and Economic Consequences of Resettlement

Case studies of other resettlement programmes would indicate that the essential prerequisites for communities to become socially and economically viable include adequate support in the initial phases which requires more than food, site, and services but also the opportunity to earn income whether in food or cash to sustain the household in the short and long term. Experience suggests that the basic aims of economic and social viability are by no means easy to achieve and even if the physical infrastructure is adequate there are a number of social and psychological factors which militate against successful resettlement. It is also common for governments and administrators to underestimate the economic complexity of even small scale subsistence communities and to assume that the mere provision of a tract of land is sufficient for viability. This is very rarely the case as has been seen in long-term refugee camps (see Holt 1981). People carry out complex economic transactions involving the sale of relief food in order to buy what are considered essentials such as spices or to capitalize a further income-earning activity. The pervasiveness of such transactions, much to the chagrin of the IDC, points to the very great need newly settled peasants have to create a kind of social security by diversifying their income sources.

Colson, in her study of resettlement of the Gwembe Tonga following the flooding of the Kariba Valley, reports that compensation money was spent on buying food 'because they had no tobacco to sell' (Colson 1971, p. 45). Colson also reports that the absence of a certain green plant with which to make green relish was a serious obstacle to the people's sense of whether or not they had a good deal from the government. The study goes on to enumerate the kind of factors which tend to affect the successful outcome of resettlement in the short term. These include on the physical infrastructural side:

- Whether or not there is suitable land and other resources to continue with a traditional and/or cash crop. In Ethiopia the capacity of the soil to grow teff would be a major consideration.
- Where services such as schools, health centres, water, latrines, and roads are, in relation to the settlements. People's perception of the purity of the water supply is of particular consequence.
- Whether cattle which represent capital on the hoof can be brought to the area and whether or not there is sufficient and suitable grazing.

Other crucial factors on the social side include who people have to neighbour and trade with, traditional enmities and tribal differences being an obvious aspect, whether habitual or traditional trading partners can be maintained, and whether traditional farming techniques can continue to be put into practice. Colson also reports that a familiarity with local flora and fauna is significant in drought-prone areas since it is normal for families to scavenge rural hinterlands for so-called famine foods which are a valuable means of tiding them over a hungry period before the new harvest is available. People also make clear and intelligent choices about how and when they will invest meagre resources. Thus a farmer may have first to be convinced that a good yield is possible in a newly settled area before risking his own precious seeds for planting.

The social consequences of large-scale relocation and resettlement programmes are more difficult to gauge. One may suppose that if the basic social and physical services were adequate and appropriate then social cohesion would automatically follow. However, it may be naïve to point out that social units are not as malleable as some planners believe or hope. The consistent theme in this chapter has been that social security is of paramount importance in combating famine and that anything which acts to disrupt social cohesion is necessarily counterproductive.

Again, the Colson study reveals that new employment opportunities profoundly affected traditional authority, marriage rules, and organized labour on fields. Men took young wives away with them to their employment centres and 'this meant that a young couple were not accumulating a debt for food and other support which could bind them to a homestead as labourers in the coming season' (1971, p. 111). Finally, Colson notes that those who were strong and successful in the aftermath of this resettlement were also those who had been strong and successful in previous village life.

Preliminary Conclusions

The outcome of the Ethiopian resettlement programme is not known other than anecdotal reports of people attempting to return to the north.

However, the scheme will have to be judged on the basis of reduced vulnerability to famine. The areas in which northern Ethiopians are being settled are undoubtedly less drought prone but the degree to which the newcomers will be willing and able to build social and economic networks will determine their viability. As has been demonstrated, famine can occur in the context of adequate or even plentiful food supplies (i.e. when there is no drought). The fact that many northern Ethiopians are now in more fertile regions does not automatically guarantee that they will not suffer acute food shortages. On the contrary, such draconian measures of social organization may weaken the myriad relationships that people have forged over generations in order to acquire some degree of social security and, in this sense, resettlement schemes may have similar effects to those of war.

DISCUSSION AND CONCLUSIONS

DISCUSSION

The position can be restated: famine prevention and avoidance is achieved through local coping-strategies which effectively maintain access or entitlement to food. These coping-responses are often rooted in traditional agricultural and other practices which support income transfer between households. Famine follows when these local coping-responses fail through sudden loss of entitlement to food. This can happen when local food shortages become acute and the poor are the first to lose agricultural wages; food prices rise making food purchase too expensive or displacement means that land and crops have to be abandoned suddenly. At this stage food relief is the only intervention that can prevent starvation.

However, given that the indicators of famine can be detected, and given that there are traditional coping-responses it should not be impossible to intervene at an earlier stage to support local coping-responses. The case studies presented in this review suggest that migrant labour as an economic option is ultimately destabilizing but also a short-term protection. The dependence of large sections of a community on cash resources which are uncertain because they are themselves subject to decision-making quite outside the recipient's control, is itself a cause of vulnerability. The second cause has to do with the social disruption to the economic unit, i.e. the insecurity that prolonged separation can generate. The third aspect of vulnerability in remittance-based economies concerns the relative lack of investment in local production that it encourages. Nevertheless, given that a remittance-based economy has become well established, in the short term it tends to make famine unlikely because remittance economies go hand-in-hand with a major food-import economy and because local drought does not necessarily cause a reduction in cash income and, therefore, in entitlement to food. Certain households, most particularly those without access to cash, will suffer but the chances of being able to raise cash are better, as has been documented for Lesotho, than in purely subsistence agricultural economies as is seen in northern Ethiopia.

Famine in Mozambique appears to result from a combination of greatly reduced agricultural production – itself following from unwise economic policies – drought, and widespread displacement as a result of war. Mozambique's vulnerability is, perhaps, greater than in either Lesotho or Ethiopia, because the social and economic networks which act as a form of security cannot operate since there is no basis for exchange and, thus, no incentive for surplus production of food grains. The greatest constraint in a country as potentially fertile as Mozambique is the lack of essential consumer goods to exchange or buy with locally produced food. Added to this, increased agricultural production can not and will not proceed until farmers are persuaded that a given area is safe from attack and, therefore, that investment is prudent.

The fact that peasants in Mozambique have not apparently starved in massive numbers must in part be due to the general fertility of the land, relative underpopulation, and the resilience of cassava to drought.

The longer-term outcome of Ethiopian resettlement programmes must await further studies. However, preliminary evidence indicates that re-settlement programmes are generally successful in economic terms if they are very heavily subsidized for some period of time and if there is a reasonable chance that traditional agricultural and economic practices can be maintained in the new locations.

Households need to accumulate tradable assets.* These, as has been noted, are more easily gained and maintained if the household or the economic unit is large and the income sources varied. Clearly these two factors are linked but the point is re-emphasized here because anything which tends to limit either the size of the economic unit or the income sources must be counter-productive. Aid inputs in the pre- and post-famine context then should help to maintain or at least allow larger households to co-operate in production and to maintain or allow varied income sources.

It is an unfortunate fact that communities that have already suffered famine are more likely to do so again in the future in spite of massive injections of aid. Although it would be unfair to suggest that the kind of aid

* Or as Smith and Christian (1984) call it, halting the 'law of accumulating disadvantages' (p. 355). Poor diets in pre-Revolutionary Russia were not merely a reflection of general poverty but actively contributed to a family's misfortune by narrowing the possibilities of economic recovery.

given causes future vulnerability, there is certainly a need for greater understanding of how far certain external assistance may discourage insurance activities. Disasters in the perception of donors and governments of countries in which they take place are, by definition, unusual events requiring an unusual response. What, however, is not explicitly acknowledged by either donors or governments is that disasters are a continuum of inadequate access to resources in order to reduce individual and group vulnerability. At times and often predictably people become so overwhelmed by a combination of events such as successive years of drought, war, unemployment, enforced migration, and/or sudden changes in the terms of trade that vulnerable groups become acutely affected. In the case of war as a cause of famine clearly the IDC cannot enforce a political solution. The case has been made, however, that the IDC at least intervene in a way that does not further destabilize the social and economic life of a vulnerable community. Free food aid may not help the community to build reserves; on the contrary it may further the gap between rich and poor by benefiting those who are already relatively well protected or by undermining mechanisms for income transfer by discouraging traditional agriculture practices.

It has been argued that free food aid, though essential in the context of starvation, does not necessarily promote recovery precisely because it does not contribute to local coping-responses. Food, other than relief for the starving, and unless it can be commuted to cash or other assets, does not shore up the individual household for future onslaughts. Three other kinds of aid can be briefly examined here.

Food-for-Work Schemes

Food-for-work programmes have a long history and in the developing world not as much success as had been hoped for. In the pre-famine context, the need is for a reliable source of food to tide households over from one season to the next. On the administrative side, there is a clear need for fair registration of the most needy, supervision of projects, and regular deliveries of food.

In addition, the development projects themselves have to be relevant in the sense that they benefit the community as a whole and are, thus, perceived as valuable. If these conditions are not met, food-for-work programmes, as famine-prevention mechanisms, will not succeed. For example, a scheme to irrigate land will not benefit the landless peasant unless he is also guaranteed agricultural wages once the project has been completed. Similarly, theoretical food entitlement by means of food-for-work projects is meaningless to the household which will starve unless it receives payment in the next week or month or whatever is the danger period. Failing this the household will have no choice but to move in search of food if food payments continue to be delayed.

The argument for expanding food-for-work programmes rapidly when food shortages are threatened is a sound one since, in principle, it *adds* an income source while not necessarily limiting existing sources however depleted they might be. Neither do food-for-work schemes disrupt social and economic units provided they are within reasonable distance of the home village. These kinds of schemes can thus be supportive by extending local coping-strategies. Moreover, the developmental impact can be considerable, as the terraced hillsides in central Ethiopia bear testimony. The problems, therefore, remain those concerned with the administration and logistics together with promoting an awareness, on the part of governments and the IDC, that resources in order to expand the schemes are fully justified on the basis of pre-famine indicators.

Food-for-work projects have a poor record in Lesotho mainly due to the fact that food wages cannot successfully compete with remittance money. However, Bryson (1984) reports that some projects have undoubtedly helped the poorer female-headed households. In Ethiopia food-for-work projects have been a great deal more successful. The question as to whether or not certain projects could have been rapidly expanded in the drought-affected areas to tide communities over and thereby prevent distress migration is an interesting one. It has been reported that existing projects, especially in the north, were vastly over-subscribed and that, perhaps, as many as 500 000 applicants would apply for 1000 places. The major problem appeared to be that there was insufficient food for rapid expansion. In 1984 there was a shortfall in both the amounts of food allocated for the programme as well as an imbalance in the commodities supplied. Moreover in the early stages of the famine, that is throughout the first 6 months of 1984, the food provided under the aegis of food-for-work projects was intended as a supplement but increasingly became the mainstay of the family's food income. At the height of the emergency, rations normally supplied within the terms of reference for food-for-work projects would have been insufficient to resuscitate the severely malnourished. In the post-famine phase the problem is to judge what *kind* of projects may help to reduce future vulnerability and, inevitably, a comparison of the longer-term economic benefits of food-for-work programmes with the government's resettlement scheme will have to be made. One further point worth emphasizing, both kinds of intervention must eventually come to an end, hopefully leaving behind a more robust community.

These administrative difficulties, therefore, predominantly those to do with the availability and delivery of food for the food-for-work projects have so far precluded any real testing of using these schemes as effective famine-prevention measures.

Cash-for-Food Projects

These innovative schemes have been implemented by Unicef in Ethiopia in famine-affected areas. In late 1983 an experimental project called Local Purchase of Food Commodities was initiated in two sites in Gonder and Shoa regions. The project aimed at providing interim relief to affected populations in the drought pocket areas by providing the destitute with cash to enable them to purchase food from surrounding surplus areas. It was anticipated that this kind of intervention because it identified and supported local opportunities and coping-mechanisms and through mobilizing the transfer of surplus production in some areas to neighbouring drought-affected areas, would achieve at least four things: reduce the delays in the delivery of assistance, eliminate transportation costs, prevent the displacement of populations in search of relief food, and encourage community development. Beneficiaries of the project who were capable of working were expected to engage in community work although the emphasis of the project was relief. Various developmental projects such as the upgrading of feeder roads, spring capping, terracing, community vegetable gardens, and latrine construction, were successfully undertaken and the results of the pilot project were sufficiently encouraging for the scheme to be implemented elsewhere.

An evaluation of the project has since been carried out (Kumar 1985) and there are undoubted short-term benefits which include the relative ease with which it was administered, the speed with which food could be transferred from surplus to deficit areas, the lower costs involved, and above all the prevention of migration of whole communities to relief camps. Additionally there are reports that the community development initiatives have been largely successful. Inevitably, however, some problems have arisen which may be more serious in the long term and limit the applicability of such schemes in the future.

According to Sen's argument (1981) a local deficit area even if surrounded by surplus-producing regions will not attract that surplus in the absence of cash. 'Market demands are not reflections of biological needs or psychological desires, but choices based on exchange entitlement relations' (p. 161). The provision of cash, therefore, would presumably cause traders to respond to the new demand. Research studies, on the other hand, have also made it abundantly clear that increased demand for a scarce resource causes inflation and possibly famine, although the failure in entitlement will be somewhat displaced geographically. The poor in the surplus area will eventually become unable to afford the available food because extra demand from the outside has artificially caused price rises. The real danger, therefore, of cash-for-food projects is that it simply shifts the famine geographically unless the scheme can become sufficiently widespread to create an overall balance between the surplus and deficit areas. Even where this is possible, however (and one very much doubts that it would be possible), there would still be pockets, possibly numerous ones, where sudden and increased demand has reduced local supplies and greatly augmented the prices, resulting in local distress. Kumar (1985) reports:

there is little doubt that cash disbursement (which immediately enables each family to purchase approximately 78 kg of cereals per month) has transformed a need for food grains into an effective demand for them: Correspondingly the increased ability of the population to command food has attracted food into the area.

However, the same author also reports that there were price rises in a neighbouring administrative district (woreda) and that local chairmen had reported concern that inflation was occurring in a previously reasonably secure surplus area. The author goes on to say that he, himself, was fairly convinced that these price rises were due to supply rather than demand factors but this is not entirely demonstrated.

These kinds of schemes are undoubtedly innovative and in certain local contexts (certainly to the recipients of cash distributed so far) may be a very valuable input; but it is suggested that the scheme is only really viable in the *pre-famine* context, when minor threats to food supply affect the very poor who are forced to start responding immediately. Attempts to stabilize the market at this stage may work. It would seem more sensible nevertheless, in spite of the obvious administrative difficulties, to meet threatened or actual food shortages with subsidized *food* rather than *cash*. Cash can cause inflation and displacement, food at subsidized prices on local markets could eliminate the immediate *causes* of famine – that is, loss of entitlement.

Labour-Based Relief Schemes

The third kind of intervention involves a combination of food-for-work programmes capable of rapid expansion in times of drought, and food distribution to the extremely vulnerable who were unable to work. The Government of Botswana has attempted such a complex programme (see Relief and Development Institute 1985). Briefly, the scheme, refined during the drought years of 1980–2, rested upon the setting-up of regional and local food-reserve depots which guaranteed food for a substantially increased number of people registered on existing food-for-work projects. In the first few years up to 20 000 people were employed for an average of 3 months and labourers were rotated to cope with oversubscription to the projects. As the programme continued a more simplified funding and approval procedure was instituted which ensured a more equitable distribution based on population numbers and the severity of the drought in the different districts. As the number of free food-aid beneficiaries continued to rise, a rationing-card system was introduced to prevent abuse.

When finally the drought began to recede, the infrastructure set up to deal with the drought relief programme was used to implement a job-creation scheme. Undoubtedly, there were many familiar administrative and logistic problems which arose in the course of this kind of approach. But its value lies in the fact that the Botswana Government was committed to the extent that it was prepared to underwrite famine containment and had an understanding as to the kind of input which would ensure entitlement to food other than through charity.

CONCLUSIONS

Sen reports that the main research interest lies not so much in proving that famine is caused by entitlement failures but in:

characterising the nature of *causes* of entitlement failures. . . . The contrast between different types of entitlement failures is important in understanding the precise causation of famine and in devising famine policies, anticipation, relief and prevention (Sen 1981, p. 164).

One of the case studies presented in this chapter (Lesotho) suggests that famine, as it is generally perceived, did not occur, because there was no breakdown in entitlement. In Mozambique, entitlement failure was and is widespread more as a result of war and displacement than of drought but massive starvation affecting whole regions was thankfully absent due most probably to a number of remaining options including the fertility of the land. The Ethiopian famine was visible and apocalyptic, and devastating for future recovery. The famine prevention measures taken by the government may or may not reduce future vulnerability. The economic fate of the resettled as well as those that have been forced to receive so many newcomers will have to be assessed.

The essential prerequisites for famine prevention include a better understanding of how, why, and when local coping-responses are beginning to falter and what inputs would support local famine avoidance-responses. The information base is perhaps the most crucial aspect because without it famine can erupt suddenly and because the IDC, in particular, cannot intervene unless and until it is fully convinced that there is a crisis to deal with. It is, therefore, the task of the research community to understand increasing vulnerability as manifested by gradual or sudden loss of entitlement to food and to document these events. Finally, if famine can be avoided by ensuring a continuance of food entitlement then earlier intervention must be geared to supporting local strategies to achieve this. In the post-famine stage, the strategy should be no different, namely an

attempt to understand and support those activities which people themselves recognize as security against future famine.

ACKNOWLEDGEMENTS

Field visits during which material for this paper was collected were carried
out under the auspices of Catholic Relief Services New York, the UN
World Food Programme, and Unicef. The author gratefully acknowledges
support but emphasizes that the views expressed are those of the author
alone and do not necessarily reflect those of the organizations involved.

The author also thanks Martin Griffiths for extensive comments on an
earlier draft of this chapter, and Professor Ainsworth Harrison for editorial
support.

LIST OF REFERENCES

Barker, A. (1973). Community of the careless. In *Outlook of a century* (eds
F. Wilson and D. Perrot), pp. 490–3. Lovedale Press & Sprocas, South Africa.

Borton, J. and Shoham, J. (1985). *Risk mapping & early warning indicators: the
Zambia case study*, Report for FAO, Nutrition Division. Rome, Italy.

Bryson, J. C. (1984). *Review of selected national experiences with food aid policies &
programmes: the Lesotho experience*, Report for WFP. Rome, Italy.

Catholic Relief Services (1983). Call forward records. CRS Office, Maseru,
Lesotho.

Cobbe, J. (1982). The changing nature of dependence: economic problems in
Lesotho. Paper given at Association of Social Anthropologists meeting. No.
82–19. Washington DC, USA.

Colson, E. (1971). *The social consequences of resettlement, Kariba studies IV*, The
Institute for African Studies, University of Zambia. Manchester University
Press.

Cutler, P. (1984). Famine forecasting: prices & peasant behaviour in northern
Ethiopia. *International Journal of Disaster Studies and Practice*, 8(1). London.

Cutler, P. and Shoham, J. (1985). *The state of disaster preparedness in the Sudan*,
Report for the International Relief & Development Institute. London.

Cutler, P. and Stephenson, R. (1984). *The state of food emergency preparedness in
Ethiopia*, Report for the International Relief & Development Institute. London.

Dando, W. A. (1980). *The geography of famine*. Edward Arnold, London.

D'Souza, F. (1984a). *The threat of famine in Afghanistan*, Report for Afghan Aid,
London.

D'Souza, F. (1984b). *The role of information in disaster relief*, Report for the
International Relief and Development Institute. London.

D'Souza, F. (1984c). The socio-economic cost of planning for hazards: an analysis
of Barkulti Village, Yasin, northern Pakistan. In *International Karakoram
Project, Vol. 2*. Cambridge University Press.

D'Souza, F. (1985a). Information and professionalism in disaster relief programmes. *International Journal of Disaster Studies and Practice, Special Supplement*, pp. 22–6. London.

D'Souza, F. (1986b). *Strengthening information networks in Mozambique*, Report for Unicef, Maputo, Mozambique.

D'Souza, F. and Shoham, J. (1985a). The spread of famine in Africa: avoiding the worst. *Third World Quarterly*, 7(3), pp. 515–31. London.

D'Souza, F. and Shoham, J. (1985b). *Evaluation of Catholic Relief Services programme in Lesotho*, Report for CRS. New York.

FAO (1975). *Report of the Food Security Policy Formulation and Project Identification Mission*. FAO, Rome.

Feacham, R. G. A., Burns, E., Cairncross, A., Cronin, S., Cross, P., Khan, M., Lamb, D., and Southall, H. (1978). *Water, health & development: an introductory evaluation*. Trimed Books. London.

Fry, M. J. C. (1974). *The Afghan economy*. E. J. Brill, Leiden, The Netherlands.

Gay, J. (1984). *Social and economic aspects of the Catholic Relief Services program to feed pre-school children in Lesotho*, Report for CRS, p. 110. Maseru, Lesotho.

Government of Lesotho (1985). Annual statistical bulletins, 1983–1985. Maseru, Lesotho. 1985.

Government of Lesotho (1985). Food and nutrition information bulletins, nos 1–4. Food and Nutrition Co-ordinating Office, Maseru, Lesotho.

Government of Mozambique. Department of Meteorological Services. Maputo, Mozambique 1980–83.

Government of Mozambique. Department for the Prevention and Relief of Natural Disasters. Annual Reports 1985. Maputo, Mozambique.

Holt, J. (1981). Social and Economic Aspects of Refugee Camps. *International Journal of Disaster Studies and Practice*, 5(3). London.

Holt, J. and Cutler, P. (1984). *Review of the early warning system of the relief and rehabilitation commission (Ethiopia)*. Report for Unicef and the RRC. Addis Ababa. Ethiopia.

Ishemo, S. L. (1978). Some aspects of the economy and society of the Zambezi Basin in the 19th & Early 20th Century. *Mozambique: Proceedings of a Seminar at the Centre of African Studies*, pp. 16–32. University of Edinburgh, December 1978.

Kumar, B. G. (1985). *The Ethiopian famine and relief measures: an analysis and evaluation*. Report for Unicef. Addis Ababa, Ethiopia.

Monet, P. (1964). *Eternal Egypt*. The New American Library, New York.

Murray, C. (1983). *Families divided: The impact of migrant labour in Lesotho*. Cambridge University Press.

Murlis, J. (1980). An analysis of food aid information in Kampuchea to March 1980. *International Journal of Disaster Studies and Practice*, 4 (3), 263–70. London.

Relief and Development Institute (1985). *Strengthening disaster preparedness in six African countries*. Report for the Ford Foundation. International Relief and Development Institute. p. 107. London.

Save the Children Fund (1983). Statistical returns. SCF Office, Maseru, Lesotho.

Seaman, J. and Holt, J. (1980). Markets and famines in the Third World. *International Journal of Disaster Studies and Practice*, 4(3). London.

Seaman, J., Holt, J. and Rivers, J. (1974). *Harerghe under drought*. Report for the Relief and Rehabilitation Commission (Ethiopia). Addis Ababa, Ethiopia.

Seaman, J., Holt, J. and Rivers, J. (1978). The effects of drought on human nutrition in an Ethiopian province. *International Journal of Epidemiology*, 7.

Sen, A. (1981). *Poverty and famines: an essay on entitlement and deprivation.* Clarendon Press, Oxford.

Shawcross, W. (1984). *The quality of mercy: Cambodia, holocaust and world conscience.* Simon and Schuster, New York.

Smith, R. E. F. and Christian, D. (1984). *Bread and salt: a social and economic history of food and drink in Russia.* Cambridge University Press.

Snow, E. (1962). *The other side of the river: Red China today.* Random House, New York.

Spiegel, A. D. (1979). *Migrant labour remmittances, the development cycle and rural differentiation in a Lesotho community.* M.A. Thesis. Dept. of Social Anthropology, University of Cape Town, South Africa.

Srivastara, R. K. and Livingstone, I. (1983). Growth and distribution: the case of Mozambique. In *Agrarian policies and rural poverty in Africa* (eds D. Ghai & S. Radwan), pp. 249–80. International Labour Organisation, Geneva.

Tobert, N. (1985). The effect of drought among the Zaghawa in Northern Darfur. *International Journal of Disaster Studies and Practice*, 9(3), pp. 213–24. London.

Unicef (1984/85). *Mozambique: annual reports*. Maputo, Mozambique.

Unicef (1985a). *Four new villages near Espungabera in Manica Province*, Survey Report. Maputo, Mozambique.

Unicef (1985b). *Accelerated rehabilitation of the food supply system and nutritional aid program to the Changara District.* Consultant's Report. Maputo, Mozambique.

Unicef. Mozambique: Annual Reports 1983–6. Unicef Offices, Maputo, Mozambique.

Wallman, S. (1969). *Take out hunger. London School of Economics Monograph on Social Anthropology*, No. 39. The Athlone Press, University of London.

Wield, D. (1978). Mine labour and peasant production in southern Mozambique. *Mozambique: Proceedings of a seminar held at the Centre for African Studies*, pp. 78–85. University of Edinburgh, December 1978.

Van der Wiel, A. C. A. (1977). *Migratory wage labour: its role in the economy of Lesotho.* Mazenod. Mazenod Book Centre, Cape Town, South Africa.

Wuyts, M. (1978). *Peasants and rural economy in Mozambique.* Discussion paper published by the Centre for African Studies. Eduardo Mondlane University, Maputo, Mozambique.

York, S. (1985). Report on a pilot project to set up a drought information network in conjunction with the Red Crescent Society in Darfur. *International Journal of Disaster Studies and Practice*, 9(3), pp. 173–9. London.

2

THE NUTRITIONAL BIOLOGY
OF FAMINE

J. P. W. Rivers

THE SOCIAL DIMENSION OF NUTRITIONAL BIOLOGY

The perception during the last decade that famine is a social phenomenon has transformed the study of the subject, and as is evident from the chapters by Dr D'Souza and Professor Desai, without this dimension famine cannot be fully understood. But if famine is a social phenomenon, it is one with biological aspects, a truly bio-social subject, and this chapter attempts to deal with part of this, the 'nutritional biology' of famine: the nature, aetiology, and implications of the nutritional diseases that characterize famine. These diseases are the most obvious aspect of our experience of famine, the grist for the television news mills, the hushed voice and horrific picture that describe the complex of diseases we call 'malnutrition'. They are also, for victim and observer, in good part the measure of suffering and the trigger for aid.

The nutritional biology of famine is not simply a study of malnutrition. But because present-day famines strike in developing countries where malnutrition is endemic, there is an unfortunate tendency to blur the differences between endemic malnutrition and famine-induced malnutrition, to treat the latter as if it were simply the former writ large. My colleagues and I protested about this after the Ethiopian famine of the early 1970s, arguing that it is vital to concentrate on the unique nature of famine. Although we wrote then primarily to argue that famine was social and not simply biological, our statements still provide the contextual basis for the present chapter:

Scientists and laymen alike write of famine as if the word were a synonym for death by starvation. In extreme cases, the definition is so stretched that phrases like 'endemic', 'chronic', or 'permanent' famine have been used to describe the nutritional manifestations of underdevelopment.

This woolly equation may make good journalism, but it makes bad science.

Unfortunately it is so ingrained that of the very few studies that have been made on famine epidemiology, most have concentrated in fact on the epidemiology of famine-induced malnutrition. This is undoubtedly an important topic, but it does not approach the real problem any more than a study of battle injuries casts light on the nature of war .... Famines are distinct. Traditional chronicles are as unequivocal about recording famines as they are in recording wars, even in a nation like Abyssinia, where poverty and malnutrition were endemic .... There is no transition between even the worst sort of normal year and the famine. They are different, not only in severity, but in kind. This is because the famine year is neither characterised by poverty, nor even death, but by social disruption. Miserable though it is, chronic poverty in traditional societies is a situation to which considerable social, psychological and physiological adaptation has occurred. Only when these mechanisms of cultural homeostasis are unable to cope does the situation shift into famine (Rivers et al. 1976).

THE NUTRITIONAL CHARACTERISTICS OF FAMINE

Three of the aspects of famine noted above account for its unique biology: the timescale of events, the nature of the nutritional deficits that occur, and the potential nutritional impact of the associated social disruption.

TIMESCALE

Clearly famine differs from underdevelopment because it is an acute, that is, short-term and severe deterioration of the nutritional status of the community from a background which may or may not have been one of chronic food shortages.

Many workers in the field have therefore accepted the hypothesis that famine can be characterized by the occurrence of an acute form of malnutrition in children, while underdevelopment is characterized by a chronic type, a view which has become central in famine epidemiology. I shall argue in this chapter that this hypothesis, although superficially attractive, is incorrect.

What distinguishes famine-induced malnutrition is not that it is *acute*, but that it is *extensive*. The aetiology of malnutrition at the family level is unchanged. But in famine the small-scale family catastrophes that precipitate malnutrition are in phase, occurring in many families at the same time, with the number and socio-economic range of families affected being increased. And above all the demographic range of cases of malnutrition is more extensive. Although in famine the vast majority of deaths are among young children and indeed the bulk of the children in a community may die, in a real sense what is different about famine is that other age groups begin to share their fate.

THE NATURE OF THE NUTRITIONAL DEFICITS OF FAMINE

Famine has often been equated with starvation, but the biology of famine is more complex than this facile view allows. Indeed relatively few famine victims are starved, in the sense of going completely without food. Most are semi-starved, and, ideally, once food aid arrives, they may not be starved at all, inasmuch as they receive a subsistence diet, which meets their minimal needs, if not their wants. Such people are none the less still famine victims; they clearly continue to suffer.

Therefore to understand the nutritional biology of famine it must be realized that it typically consists of two superimposed phenomena, some degree of food shortage and the consumption of a strange diet. This diet may comprise food items of which the individual and his community have little previous experience: famine foods, such as wild plants, traditionally eaten when food is short, or food aid, or, *in extremis*, such things as rats or even dead bodies. Or it may consist of well-known and acceptable foods eaten in unusual amounts: cereal grains or starchy roots come to dominate the famine victim's diet to a degree unusual in normal times, not least because much famine aid is cereal grains, as Table 2.1 illustrates. These famine diets are low in energy, contain normal levels of protein and very low levels of fat, and hence have a high content of carbohydrate.

The dietary deviations of famine have received surprisingly little scientific attention, and yet they are most important in dictating the pattern of nutritional disease that occurs.

SOCIAL DISRUPTION

The social disruption of famine, described by D'Souza (1987) has a profound nutritional impact. It arises from the poverty that is, in effect, the *sine qua non* of famine. Famine victims are pauperized, and frequently have sold belongings, including clothing, to obtain food. They may be unable to buy fuel, or have lost their homes, or, often driven by their poverty, have become refugees. It will be shown below that such social disruption can have a profound influence on nutritional needs.

LITERATURE SOURCES

Famines are not ideal situations for nutrition research, so that the reviewer is forced to draw evidence from a variety of situations which mimic pure

TABLE 2.1. Nutritional characteristics of some famine rations

Famine and date	Energy kcal/day	Protein	Fat	% of Energy as: Carbohydrate	Cereals	Sugar	Notes
Belsen Concentration Camp, 1944–5	~400						1, 5, 8
Warsaw Ghetto, 1942–3	800	10–15	3	>80			5, 9
Dachau Concentration Camp, Sept. 1944	1020	~10	<10	~80	59	5	1, 5, 10
Dachau Concentration Camp, Apr. 1945	530	<10	<3	~90	45	0	1, 5, 10
POW Camp, Tost, Germany, 1939–45	1610	14	15	71	52	6	2, 4, 10
Parisian Insane Asylum, 1941	1700	15	13	72			4, 10
Changi POW Camp, 1942–5	2640	11	14	75	64	3	1, 2, 4, 6, 11
Hong Kong Internment Camp, 1942–5	2010	11	18	~70			1, 3, 5, 11
Singapore Internment Camp, 1942–5	2060	13	16	~71	68	3	1, 3, 11
Somalia, Hira Refugee Camp, 1980	1050	12			49	0	12
Somalia, Quorioley Camp, 1980	1390	9	13	~78	83	5	12
French POW's, Germany, WWI	2245	~13			77	5	10
Russian POW's, Germany, 1942–5	1600	15					7, 14
Emergency subsistence diet	1200	10	18–20		81		7, 14
Temporary maintenance diet	1500	11					13

Notes and Sources: Blank spaces denote that no data were available.

(1) Theoretical rations. Inmates state that rations were often below this.
(2) In addition Red Cross parcels were received by inmates.
(3) Includes contribution of Red Cross parcels.
(4) Population of male adults only.
(5) Population of adults and children.
(6) Population doing forced labour.
(7) Proposed ration scales for mixed populations.

(8) From Collis (1945).
(9) From Apfelbaum (1946).
(10) From Keys *et al.* (1950).
(11) From Smith and Woodruff (1951).
(12) From Tresalt *et al.* (1985).
(13) From Leyton (1946).
(14) From Masefield (1971).

famine in some respect. Two of these have proved especially valuable and are frequently cited in this chapter.

The first are studies conducted on the half-starved populations interned by the German and Japanese authorities during World War II in prisoner of war (POW) camps, and, particularly, in concentration camps. The captors collected little data, the liberators some, but quite remarkably, scientists and physicians in the prison populations often undertook quite extensive studies of their conditions. As well as individual survivors who reported their observations (e.g. Adelsberger 1946), there were two magnificent group studies. A deeply moving study published as *Maladie de Famine* (Apfelbaum 1946; Winick 1979) was made by the Jewish physicians in the Warsaw Ghetto during the 7 months before its liquidation in 1943. A similarly impressive monograph resulted from the painstaking study of nutritional deficiency diseases undertaken by British physicians in Japanese internment camps (Smith and Woodruff 1951). I pay tribute to the fortitude of these observers and the merit of their work.

The second major source is the few experimental studies of semi-starvation in humans, of which Benedict's (1915) study was a pioneering effort and *The Biology of Human Starvation* (Keys *et al.* 1950) the most thorough. This latter is a study of underfeeding American conscientious objectors for 6 months in World War II, accompanied by an exhaustive review of the literature on all aspects of famine, and from it I have frequently drawn. It remains the definitive work on the topic.

THE BIOLOGY OF HUMAN STARVATION

Catabolism never ceases so that all animals require a supply of nutrients to replace the obligatory catabolic losses. Nutritionists call this the 'maintenance requirement'.* When food supplies are inadequate to meet this, the deficit comes from the tissues of the body. If the deficiency is of protein or energy, this leads to tissue breakdown, and weight is therefore lost. Where the shortfall is of micronutrients, tissue levels of nutrients fall, and though tissue mass may be maintained, function is impaired, something manifest clinically as a nutrient deficiency disease.

In addition to this maintenance need, growing animals also require food for growth; consequently their nutritional vulnerability is higher. Growth does not impose a particular burden on micronutrient or even protein requirements (less than 0·2 g of protein are needed for every gram of weight gain). However, the energy cost of growth is quite high: energy intake must

* The maintenance requirement is more precisely defined as the obligatory losses plus the costs of replacing those losses, including such factors as the expenditure on consuming food, and the minimal activities associated with subsistence (FAO/WHO/UNU 1985).

be increased by about 25 kcal to retain one gram of protein, or about 5–6 kcal per gram of weight gain (Payne 1972; Pacey and Payne 1985; FAO/WHO 1973).*

The costs of growth per unit weight gain appear to be constant for all species (Payne 1972). But as humans are the slowest-growing mammals, the *demands* of growth are lowest in our species, and the differential nutritional vulnerability due to growth correspondingly lower, something that must be borne in mind when interpreting animal experiments on undernutrition.

Growth is quickly inhibited by a shortfall in energy and possibly also protein intake, and an early sign of food shortage in children is therefore a reduced growth rate. It is unclear whether deficiencies in other nutrients directly inhibit growth, or whether the growth failure in micronutrient deficiency is, like that in many infectious diseases, secondary to anorexia that occurs. Because of the presumed aetiology of reduced growth, or growth failure, it is often described by nutritionists as Protein-Energy Malnutrition (PEM). The name has been ratified by its common usage, but as an aetiological description is misleading; in most circumstances it is a result of inadequate food intake (i.e. it is 'Energy-Malnutrition'). Specific-protein deficiency almost never occurs (McLaren 1974).

When growth is impaired, in famines as in other situations, morbidity and mortality are increased, particularly in younger children. Therefore much attention has been focused on PEM in young children, particularly during the last 40 years, and an extensive literature exists on the topic. But, as argued above, PEM in children is the same disease whether it was acquired in a normal underdeveloped country or in a famine. The only real differences relate to prevalence and diagnosis, and these are the only aspects that I shall deal with at any length. Equally, because impaired growth in older children, or weight loss in adults, is rare outside famine, and consequently underdiscussed, I shall concentrate on this.

LES MALADIES DE FAMINE

If nothing whatever is consumed, the limiting factor in survival is water. Death from dehydration occurs in a few days when water losses are high, as in a hot environment, at high altitude, or when sweating is profuse, and always before any other nutritional deficiency appears. Primary dehydration, due to inadequate intake, has rarely been observed among famine victims, not least because those without water are unlikely to reach a relief

---

* These figures are compatible given that lean tissue contains about 18–20 per cent protein and 80 per cent water, and that during growth, tissue gain is on average 80 per cent lean and 20 per cent fat (Payne 1972).

feeding centre where most observers are. Secondary dehydration as a result of diarrhoea is a more likely cause of death. In both cases the clinical signs are the same: the subject is weak and apathetic with sunken features, oliguria is present, the eyes are sunk into the orbits, the skin is loose and inelastic, and when pinched stands up away from the subcutaneous tissue. In severe dehydration the presentation is one of oligaemic shock, with severe vasoconstriction, high pulse rate, and low blood-pressure (Robson 1972).

If water is available, but no food consumed, death may not come for three months or more, depending upon initial body fat. The cause of death in total starvation is obscure; it may be due to electrolyte depletion, especially magnesium depletion (Keys *et al.* 1950), or it may be due to loss of vital body protein particularly heart muscle (Widdowson 1985).

Apart from hunger-strikers, such total deprivation situations are rare. In famines, the usual situation is that water and some food are available, albeit intermittently. In such circumstances the clinical outcome is more difficult to predict, being essentially a race between energy deficiency and various vitamin and mineral deficiencies. The actual result depends upon the amount and composition of the food consumed, and the body stores, and rates of catabolism of each nutrient. The times taken for different nutrient deficiencies to appear is very variable as the estimates in Table 2.2 show. This is not to say that the outcome is unpredictable. Far from it, some nutrient deficiencies have not been reported in famines, while others are extremely common, despite the variety of famine dietaries.

In the past, relief agencies seem to have oscillated between believing that there is a risk of any and every deficiency, and concentrating on very few, notably those of protein, vitamin A and vitamin D, none of which is actually a very likely result of famine. An erroneous belief in protein deficiency in particular shaped famine relief and development aid until the late 1970s and probably cost hundreds of thousands of lives. Vociferous objections that energy not protein deficiency was the more important eventually led to a change in aid policy, stressing the adequacy of ordinary foods such as cereals.

However, the pendulum may have swung too far. In the current Ethiopian famine, as in Somalia in 1982 (Magan, Warsame, Ali-Salid, & Toole, 1983), there have been horrifying scurvy outbreaks in the camps of Somalia and southern Sudan which have occurred principally because attention was so focused on the need to supply energy as cereal grain that there was failure to appreciate that the diet was deficient in vitamin C.

With hindsight, scurvy was all too predictable and for this reason alone it is probably worthwhile recording the vitamin deficiencies that have been associated with previous famines. Despite the variety of diets upon which different famine victims subsisted, the list is relatively short and quite distinct from the textbook catalogues of nutrient deficiencies.

TABLE 2.2. Vitamin depletion times*

Vitamin	Experimental depletion adequate energy consumed	Time for appearance of disease in famine
Retinol (A)	>18 months (1, 2)	–
Thiamin (B1)	5–43 weeks (3)	12 weeks (4)
Riboflavin (B2)	13–19 weeks (5)	16 weeks (4)
Pantothenic acid	>6 weeks (6)	3–8 months (7, 14)
Folic acid	–	<6 months (4)
Niacin, normal diet	>3 months (7)	3 months (4)
Niacin, mixed diet	7 weeks (8)	~2 months (9)
Ascorbic acid (C)	8–13 weeks (10)	3–6 months (11)
Cholecalciferol (D)	30 weeks (12)	–
Essential fatty acids	>6 months (13)	–

* Times taken for the appearance of clinical deficiency disease in subjects consuming vitamin-free diets, compared with reported incubation periods of such disease in famines. Figures in brackets indicate source of data for estimate. – indicates no data.

Sources: (1) Hume & Krebs (1949).   (8) Goldsmith *et al*. (1952).
(2) Vitamin A subcommittee (1945).   (9) Author's estimate, various sources.
(3) Sinclair (1982).   (10) Carpenter (1986).
(4) Smith and Woodruff (1951).   (11) Seaman pers. Comm.
(5) Sebrell and Butler (1938).   (12) Sinclair (1982).
(6) Jeffrey (1982).   (13) Rivers and Frankel (1981).
(7) Najjar *et al*. (1946).   (14) Glusman (1947).

These qualitative deficiencies are superimposed upon the central Maladie de Famine, the wasting and variety of pathological symptoms, variously described as hunger disease, cachexia, or even PEM in adults. Most of the symptoms can be attributed to energy deficiency, but the extent to which it is complicated by protein deficiency or as yet unrecognized nutrient deficiencies is a matter of conjecture.

QUALITATIVE NUTRITIONAL DEFICIENCIES IN FAMINE

VITAMIN C DEFICIENCY

Even though the latest outbreak seems to have been greeted with such surprise, vitamin C deficiency, scurvy, is in fact a frequent accompaniment of famine. It was widespread in the Irish famine (Curran 1847; Shapter 1847; Carpenter 1986), where the local name for the disease, *blackleg*, was apparently a reference to limb colouration due to extensive subcutaneous haemorrhages (Woodham-Smith 1968). Ironically it was not at first recognized by the physicians who had hitherto associated it totally with sea-voyages, although a consensus was eventually established that the cause was the high cereal diet sent as aid, and that the inclusion of vegetables or potatoes provided both prophylaxis and cure (Carpenter 1986). Scurvy has also long been known as a problem of poorly provisioned camp populations: for example POW camps from the American Civil War to World War I have been badly affected by the disease (Carpenter 1986). These lessons of history were too easily forgotten.

Scurvy is a recognized risk in sieges. In the Siege of Paris which lasted just 4 months from September 1870, it was recorded that 'The investment of Paris and the state of want which resulted, provoked an explosion of scurvy and at the same time, less well-defined states of debility that were even more serious' (Lasegue and Legroux 1871); cited by Carpenter 1920, p. 124). Another massive epidemic was associated with the siege of Port Arthur in the Russo–Japanese war; by the end of the 9-month siege, 50 per cent of the 17 000 strong garrison had succumbed to scurvy (Hess 1920).

Like the Sudan outbreak the primary cause of all these scurvy epidemics is not food shortage as such, but due to having subsisted on 'dry rations', predominantly cereals, diets which by their nature do not contain much vitamin C (although if the grain were allowed to germinate, its vitamin C level rises to quite high levels (Passmore and Eastwood 1986)).

Scurvy is not more consistently associated with famine, primarily because the roots, leaves, and tubers which are often famine foods (Rivers et al. 1974), are themselves good sources of vitamin C. Thus, the doctors in the Warsaw Ghetto specifically note the absence of scurvy (Apfelbaum 1946) as did the physicians in Japanese camps (Smith and Woodruff 1951), in both cases despite prolonged subsistence on a minimal diet. In the former case vitamin C was probably consumed in tubers, in the latter in leaves. The effectiveness of wild foods is most eloquently underlined by the Crimean War, where in the first year, scurvy was a great problem amongst the British troops who overwintered on then standard British army foods of meat and biscuit, but not amongst their French allies who, though similarly provisioned, were ordered to scour the countryside for such items as herbs and dandelion leaves to add to their diet (Drummond and Wilbraham 1957; Carpenter 1986).

Scurvy is one of the more rapidly developing deficiencies that might be expected in a famine (see Table 2.2) though it still takes weeks to appear. This is because, while there are no reserves of vitamin C analogous to those of vitamin A, the normal tissue level of 1000–1500 mg (Hodges et al. 1971; Baker et al. 1971) takes a considerable time to deplete. In a classic study on conscientious objectors in World War II, it took at least 120 days of a vitamin C-deficient diet before clinical signs of scurvy were evident (Bartley et al. 1953). The most rapid depletion-times recorded, by Hodges *et al.* (1969), were not much less: follicular keratosis appeared after 8–13 weeks of depletion and gum signs in 11–19 weeks. These results not only demonstrate that scurvy is therefore unlikely to be a problem in famines of less than 60 days standing, but also that effective prophylaxis could be achieved even if vitamin C sources were only intermittently available. (Magan *et al.* 1983).

Keys *et al.* (1950) suggested that depletion might take even longer in an underfed subject whose turnover of vitamin C would be slower as a consequence of a reduced metabolic rate. Moreover, as Table 2.3 shows that considerable amounts of the body's ascorbic acid is in the liver and muscle and as tissue from these organs is lost in starvation, it is likely not only that the ascorbic acid they contain is made available to other tissues, but that the requirement is reduced.

TABLE 2.3. Ascorbic acid content of body organs and tissues in normal subjects, and estimates of maximal values after 40 per cent body weight loss

	Normal (mg)	Estimate after 40% weight loss*
Total	1500	810
Skeletal muscle	560	290
Brain	380	290
Liver	220	80
Kidneys	30	20
Pancreas	30	15
Lungs	30	20
Spleen	20	10
Blood	20	10
Heart	15	10
Total of organs	1305	745

* Normal value adjusted for body weight 70 kg. Loss in organ mass as shown in Table 2.4, and assuming that 15% of weight of organs is increased volume of ECF, with no necessary change in ascorbic acid content of tissue except that due to rise in ECF.

Source of normal values: Data of Yarvarsky, Almaden and King (1934).

There are still unresolved mysteries about scurvy in famine. For example, Dr John Seaman of the Save the Children Fund (UK) informs me that while in the 1986 famine scurvy was rife amongst the Ethiopian refugees in camps in Sudan and Somalia, it did not occur in the populations in camps in Ethiopia. Yet the population, their history, and their diets are apparently identical, the only obvious difference being that it is colder in Ethiopia, though it is difficult to see why this should be important.

VITAMIN A DEFICIENCY

Although vitamin A deficiency has historically been associated with famine it is probably unreasonable to regard it as a famine condition at all, except possibly for infants and young children. Much of the evidence suggesting deficiency in famine must be regarded as suspect, and in so far as it is coexistent with famine it is largely a reflection of pre-existing endemic deficiencies in areas where famine occurs. For example, there was no evidence of vitamin A deficiency in the Warsaw Ghetto, despite the fact that intakes of the vitamin were negligible (Fliederbaum *et al.* 1946; Fajgenblat 1946; Winick 1979).

Reports of vitamin A deficiency as a *de novo* problem in famines chiefly rest on two pieces of evidence. The first is night blindness, which has been reported to have been a problem in a number of famines (Keys *et al.* 1950). Although night blindness is an early manifestation of vitamin A deficiency, Keys concluded that the balance of the evidence overall is that it has a different and as yet unexplained aetiology in famine. One curious aspect of famine nyctoplia is that it is much more prevalent in men than women, up to tenfold differences being reported (Keys *et al.* 1950). This is the converse of what might be expected if it were vitamin A related. The skin-changes observed in famine victims have also been attributed to avitaminosis A by various authors. However, the details of the skin changes are not fully compatible with avitaminosis A, both scurvy and essential fatty acid deficiency as well as poor hygiene being more probable causes (Keys *et al.* 1950; Winick 1979).

Previously well-fed populations kept under conditions of food shortage and with very little vitamin A in their diet do not become deficient except in the extremely long term (Table 2.2). The major reason is that vitamin A can be stored in the body apparently without limit, even, indeed, up to toxic levels (Moore 1957). In the well-nourished adult the body stores are sufficient to provide vitamin A needs for a considerable period on a diet free of the vitamin. For example, post-mortem studies of 40 adults dying in accidents in the UK during the period 1931–5, that is prior to the advent of statutory vitamin A supplementation of margarines, showed that the median vitamin A level in liver was 220 international units (66 mcg retinol equivalent (RE)) per gram (Moore 1957). For a liver size of 1500 g this is sufficient to meet the minimal vitamin A requirement for about a year. Later studies, after the advent of fortification, showed vitamin A levels were nearly twice as high, and studies elsewhere in Europe and the USA confirmed the existence of these large liver stores (Moore 1957). The efficacy of these stores was confirmed by a study in which British conscientious objectors consumed a vitamin A-deficient diet. Only after 400–500 days were even the earliest signs of deficiency observed, and then in only some subjects (Hume and Krebs 1949).

However, it would be wrong to become complacent about this. Investigators in the USA and Canada have found that although mean liver vitamin A stores might be adequate, the variability is very great and in 20–30 per cent of individuals the liver levels are so low ($< 40 \, mcg/g$ liver) as to provide a negligible store (Hoppner *et al*. 1968; Raica *et al*. 1972). In Bangladesh it has been suggested that 78 per cent of adults may be in this category (Abedin *et al*. 1976). If a minimal requirement of $440 \, mcg/day$ is assumed (Hume and Krebs, 1949), this would only provide protection for about 4 months. Quite why such subjects do not rapidly become vitamin A-deficient in famine is not clear to me, unless their lower liver stores reflect lower requirements.

Relative to their requirements, children's vitamin A stores are lower than adults', but even here post-mortem studies suggest that median stores may be adequate for at least 150 days (Moore 1957). Infants however have very low stores. The baby is born with only trace levels of the vitamin in its liver, and since levels in the mother's milk will vary with her intake, the infant and younger child are particularly at risk in famine (Oomen 1976). Perhaps more vitamin A deficiency would have been observed in the Warsaw Ghetto had the severity of starvation not meant that fecundity was reduced and infant survival very low.

The vitamin A intake in famine will usually be low, although populations who keep food animals might need to eat them as their condition deteriorates, as at the start of a drought, and may therefore increase their vitamin A status. Similarly the consumption of much of the leafy vegetable matter that makes up famine foods will provide further vitamin A as its precursor, β-carotene, and intakes may become so high that carotene pigment causes a yellow discolouration of the skin (Levin 1944; Musselman 1945; Keys *et al*. 1950). Preformed vitamin A does not cause skin colouration; it is, however, toxic. The only poisoning case attributable to famine that I have identified is Rodahl's (1949) report of a starving Eskimo who in desperation ate the liver of a polar bear; polar bear liver contains toxic levels of vitamin A and Eskimo tradition has it that it is a toxic food. However, vitamin A poisoning, presumably non-fatal, has been associated with famine relief, as a result of a somewhat over-enthusiastic attempt at mega-vitamin A dosage (200 000 international units, 60 mg retinol equivalent (RE)) attempted by UNICEF as a prophylaxis against deficiency (Oomen 1976).

It is sometimes stated that vitamin A deficiency could be induced in famine by impaired absorption, particularly of β-carotene, because the diet is low in fat (Winick 1979). Absorption is thought to be impaired when the fat level is below 5 per cent of the energy (Passmore and Eastwood 1986), but famine diets generally contain more fat than this (Table 2.1) and I know of no instance where the problem has actually been shown to have occurred in a famine.

BERIBERI

Depletion of thiamin stores in the body has been experimentally shown to be potentially very rapid (Table 2.2). In famine, even when the intake of thiamin is low, the appearance of beriberi is dependent also upon dietary balance and levels of energy expenditure. Beriberi is associated with diets which, besides being low in thiamin, also provide a high fraction of energy as carbohydrate (Thurnham 1978), which as Table 2.1 shows is characteristic of many famine diets. Since cereal germ is the major source of thiamin in diets, diets which pose a special risk are those dominated by white (milled) cereals, including rice, or cassava, or tubers. The thiamin requirement appears to vary with the level of energy expenditure, so that famine victims will be less susceptible if underfed, and more at risk if work outputs are high. Work of course takes many forms; there is, for example, a report that claims a beriberi outbreak in a camp for Cambodian refugees in 1978, where the problem was confined to members of the camp football team.

Smith and Woodruff (1951) provide the definitive report on famine beriberi in their study of allied POWs of the Japanese during World War II, whose meagre rations consisted chiefly of white rice and whose mean thiamin intake was estimated to have been 0·4 mg per day. Beriberi began to appear quickly, the wet form being evident within 6–8 weeks of internment. At its peak, in June 1942, 4 months after internment, the weekly incidence of subacute wet beriberi reached 3·5 per cent in the Changi POW camp. In the civilian camp the disease first appeared 12 weeks after internment, and the average weekly incidence for the period March to July 1942 was 0·34 per cent. Intermittent prophylactic supplementation thereafter reduced the rate, but when supplies of prophylactic foods were not available the disease reappeared with the incidence being inversely proportional to the ratio of thiamin to carbohydrate energy in the diet (see Fig. 2.1).

Beriberi also occurred in the sieges of Leningrad (1943) and Kut el Amara in 1916 (Hehir 1922). In the latter, the British soldiers in the besieged garrison subsisted on a diet predominantly based on white wheat flour. After 5 months they were badly affected by beriberi while Indian Sepoys of the same battalion who consumed wholewheat chapattis were not affected (Passmore and Eastwood 1986), although peculiarities of the Sepoy diet led to their succumbing to scurvy (Carpenter 1986). Beriberi was also reported in isolated fishing communities in pre-war Labrador and Newfoundland cut off during the winter and subsisting predominantly on cereals (Ackroyd 1930). There was, however, no thiamin deficiency in the Warsaw Ghetto (Apfelbaum 1946) probably because the black bread they were supplied with would have been wholegrain rye.

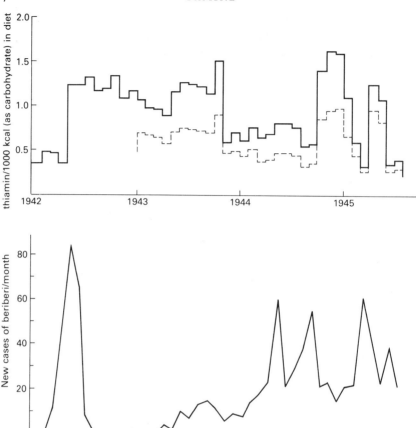

Fig. 2.1. Incidence of beriberi among British prisoners of the Japanese in the Hong Kong Civilian Camp and levels of thiamin in the diet. Thiamin levels as mg/1000 kcal of carbohydrate. For diets, continuous line shows normal rations; dotted line shows rations for persons engaged in heavy work. From Smith and Woodruff (1951).

In many countries, for example the UK, all white flour must by law be supplemented with thiamin, to avoid any risk of deficiency on diets dominated by this (Passmore and Eastwood, 1986). However, national practices vary and there is no guarantee that flours used for food aid will be supplemented, nor that whole grains will be fed as milled to produce high extraction or supplemented flour, nor that rice will be parboiled.

There is probably much more to be discovered about the relationship of beriberi to famine. There are at least three distinct forms of beriberi: wet, or cardiac, beriberi; dry, or neuritic, beriberi; and cerebral beriberi, or the Wernicke–Korsakoff syndrome. Wet beriberi is associated with oedema, indeed it was at one time thought to be the cause of famine oedema (see McCance 1951) but this idea has since been discarded since the presentation and predisposing factors are different (Beattie *et al.* 1948; McCance 1951; Keys *et al.* 1950).

Wet beriberi was recorded in the Japanese internment camps as an acute disease. The development of the disease from the first observed signs was rapid (3–7 days), typically following bacillary dysentery. The prognosis was poor. Wet beriberi also occurred in subjects with quite high thiamin intakes (0·4–0·8 mg per 1000 kcal) relative to levels found in laboratory studies in humans sufficient to prevent the disease (values of 0·23 mg per 1000 kcal were found adequate by Keys *et al.* (1950)). Smith and Woodruff (1951) suggested that factors such as diarrhoea may have impaired intestinal synthesis of the vitamin or that there was some anti-nutritional factor in rice, a suggestion which finds some confirmation in laboratory experiments.

Dry beriberi, a peripheral neuritis, was also found in the Japanese POW and internment camps (Smith 1947; Smith and Woodruff 1951). The first cases developed somewhat sooner than did wet beriberi, the first patient developing symptoms after only 28 days internment, but the progression of symptoms was slower, and recovery when thiamin was fed was also very slow. Dry beriberi was marginally more prevalent overall than the wet form, although it is a less obvious disease. Smith and Woodruff note that it was not clear how far the peripheral neuropathies observed were due in part or wholly to associated deficiencies in other B-vitamins, particularly riboflavin, something which still has to be resolved. It is, however, clear from their descriptions that the signs and symptoms of the disease were sufficiently vague that they could quite easily have been missed by less astute observers.

The Wernicke–Korsakoff syndrome presents as defects in memory and learning, but on *post-mortem* is found to be associated with considerable degeneration of the CNS (Thurnham 1978). This was identified on clinical and *post-mortem* evidence in 52 of the deaths from dry beriberi in the Japanese camps (De Wardener and Lennox 1947). Its presence is somewhat paradoxical. Although undoubtedly due to thiamin deficiency, outside disasters it is usually associated with alcoholism (Thurnham 1978) and Vedder (1913) in the classic study of endemic beriberi in the Far East makes no mention of its presence under normal circumstances. The possibility that there was something else about the rations that the population subsisted on in the Japanese camps which precipitated the symptoms must be borne in mind. De Wardener and Lennox (1947) thought it was associated with a particularly severe deprivation of thiamin and had some relationship with dysentery. There is no information as to whether it is present in other famine-afflicted populations, and without *post-mortems* it is likely to be very difficult to differentiate it from the general psychological deterioration of severe starvation.

PELLAGRA

Although pellagra can occur in famines as a consequence of simple niacin deficiency, this is rare. The only certain cases I know of occurred in the extreme conditions of Auschwitz (Adelsberger 1946) and some of the Japanese camps (Lewis and Musselman 1946; Smith and Woodruff 1951). Prison camp populations may be prone: pellagrous symptoms were reported amongst many groups of prisoners in World War I, not least Turkish prisoners of the British (Keys *et al.* 1950), but it is not certain if these were simple deficiencies. I have located no reports of pellagra from the European famines, nor from the recent African famines. This distinction has long been noted. Lussana and Frua wrote in 1856: '[In] human history ... far too often there have been great famines, destruction of whole hungry cities, penury and inadequacy of food supplies ... with the endless consequences of diseases and deaths, with epidemics ... but without pellagra' (cited from the translation by Ferro-Luzzi reproduced in Carpenter (1981)).

There is, however, a risk of pellagra in famine not due to simple deficiency, but as a result of famine relief, depending on the cereal used. Although cereals contain niacin, the availability of the vitamin from cereal sources is low (<30 per cent (Carpenter and Lewin 1985)), due to a binding protein that limits absorption. As is well known, humans can synthesize niacin *in vivo* from the essential amino acid tryptophan, and on most high cereal diets it is tryptophan that contributes most to the niacin supply.

Historically, pellagra has been most frequently associated with diets high in maize (for a review see Carpenter (1981)), which is unusual for a cereal in that it contains very little tryptophan and therefore the consumer is reliant on the limited amount of poorly available preformed niacin it contains. Where maize is used as the primary relief grain, as it is currently in Somalia, there is a risk of the disease. It was the enforced consumption of maize in the economic depression of the 1920s that caused the pellagra epidemic amongst the poor of the southern states of the USA (though it is possible that inadequate riboflavin intakes were also associated with the development of the disease observed (Carpenter and Lewin 1985)). The new genetic variant of maize Opaque-2, contains a high level of tryptophan and so presents no special risk of pellagra beyond that of other cereals (Passmore and Eastwood 1986). However, I do not know if this is a major type of maize used in food aid for famine relief.

Maize is not the whole explanation for pellagra, and among the other factors involved are excessive levels of the amino acid leucine, and dietary mycotoxins (Carpenter and Lewin 1985). The former is blamed for the occasional occurrence of pellagra amongst populations consuming ragi, an Indian millet (Passmore and Eastwood 1986). The latter is a risk in fungally infested cereals that might normally be discarded but get consumed in famine.

The relationship of pellagra to famine is illustrated by what happened in the Japanese internment and POW camps (De Wijn 1947; Smith and Woodruff 1951). In the Hong Kong internment camp, doctors reported that pellagra occurred as a result of consuming a rice diet, deficient in niacin, with symptoms developing with a 2–3-month time lag when niacin intake fell below 8 mg per day, and disappearing when it was above 10 mg. The disease, however, was relatively mild, being confined to an oral dermatitis and some skin changes (Smith 1947; Smith and Woodruff 1951).

In Singapore, however, the population subsisted for 4 months on a diet that provided only 6·3 mg of niacin per day with no sign of the disease, and then for a year on a diet which provided 15 mg niacin per day while they again remained disease-free. Pellagrous symptoms only began to appear in April 1944 when, after the niacin intake had fallen to below 7 mg for 7 months, the neutral protecting power provided maize to supplement the meagre rations. The incidence of the disease began to rise before the maize arrived, but the extent of the increase and the severity of the presentation were both exacerbated by the inclusion of maize. At this point even intakes of 20 mg niacin per day were unable to get it under control. A similar outbreak occurred in the Changi POW camp when maize was incorporated into the diet at a level of 130 g per person per day. This outbreak was complicated by the fact that the pellagrins exhibited an adverse, supposedly allergic, reaction to black beans (*Centrosoma pubescens*) that were included in the camp diet. Since local tradition had it that these beans should not be consumed, it is possible that the pellagrins were hypersensitive to the toxic effects of these beans (Smith and Woodruff 1951).

ESSENTIAL FATTY ACID DEFICIENCY

This is a deficiency which is generally regarded as rare but which Winick (1979) has suggested may have more significance in famine than has hitherto been realized.

Usually, it is held unlikely to be a problem in semi-starvation as long as weight loss continues even if the diet provides no fat (Rivers and Frankel 1981). This is because there are considerable amounts of essential fatty acid (EFA) in adipose tissue triglycerides and as this is mobilized to meet energy needs, so essential fatty acids are supplied to the body. In addition the membranes of all cells are rich in EFA so that as lean tissue is lost EFA are again made available to the general metabolic pool. However Winick (1979) has pointed out that in the Warsaw Ghetto, where the rations provided less than 5 per cent of the energy as fat (see Table 2.1), the skin changes, which observers could not explain by known vitamin deficiencies, were compatible with EFA deficiency. The diets in the ghetto were so extremely low in fat, that it is doubtful if the diet as such would have provided the 1 per cent of energy needed to meet the minimal EFA requirement.

Winick's suggestion is plausible and might merit investigation in future famines. In its favour it must be admitted that a number of the symptoms observed in famine victims, e.g. infertility, failure of wound healing, and disturbed water balance, are also compatible with EFA deficiency (Rivers and Frankel 1981). However, their low metabolic rates are incompatible with EFA deficiency (Rivers and Frankel, 1981). Moreover, a recent study of biochemical signs of EFA deficiency amongst patients with anorexia nervosa does not provide support for the theory (Langan and Farell 1985). These extremely emaciated subjects showed biochemical changes compatible with low EFA intakes, but nothing that could be interpreted as evidence of deficiency. At present, therefore, I remain unconvinced about its practical importance in famine.

RICKETS AND OSTEOMALACIA

Rickets, vitamin D deficiency in children, is often assumed to be a famine problem (de Ville de Goyet *et al.* 1978), but it probably is not. It is certainly a problem in populations whose food supply is restricted in kind (Passmore and Eastwood 1986). But, although it is sometimes observed in famine-afflicted populations, the balance of the evidence from famine is that it does not occur as a new problem. For example in the Warsaw Ghetto its absence was specifically remarked upon, though there was some radiographic evidence of sub-clinical bone changes (Fliederbaum *et al.* 1946). Although it is commonly supposed to have been a problem in Europe during the food shortages following World War I (Vincent 1985), it is probable that the reports ignored the pre-famine pattern of disease, and no real increase occurred (Keys *et al.* 1950). Indeed, Simon (1919) commented on the low prevalence of rickets in malnourished children in Europe at the end of World War I suggesting that it was the fact that such children were too weak to stand, and were carried by their mothers that protected them against the disease, which is normally due to distortion of the bones unable to support body weight.

There are various factors that would serve to militate against rickets in famine. One is that such populations are frequently exposed to the sunlight because they are refugees and have lost or sold much clothing, and vitamin D can be synthesized in the skin, provided the level of ultraviolet irradiation is sufficient. The second possible explanation is that rickets is a disease of the growing child where bone growth continues without mineralization. If growth has been arrested the vitamin D needs become correspondingly less important (Fraser 1981). Starvation indeed may have a beneficial effect on pre-existing rickets: McCollum *et al.* (1922) showed in rats that starvation was associated with the healing of pre-existing rickets.

Although rickets may be rare in famines, osteomalacia, bone demineralization in adults, is not, and the syndrome of 'hunger (or famine) osteomalacia' is well described in the classic literature (see, e.g. Pompen *et al.* 1946; Fliederbaum *et al.* 1946; Keys *et al.* 1950). I have found no information on the condition in the literature on recent famines: it may be that it is a purely European phenomenon, or it may be that it is not currently being looked for.

Vitamin D deficiency may play some part in hunger osteomalacias, but other factors are also involved. Winick (1979) suggested magnesium deficiency was involved in the osteomalacia in the Warsaw Ghetto. A more likely primary cause is mobilization of bone minerals secondary to a metabolic acidosis that occurs in starvation. Nevertheless it is of interest that in the 24 cases of hunger osteomalacia identified by Pompen *et al.* (1946) in wartime Holland a significant fraction occurred in nuns, who would have less skin exposure to sunlight, and therefore a lower vitamin D status.

RIBOFLAVIN DEFICIENCY

Theoretically, riboflavin deficiency should be a likely deficiency of famine, since the major dietary sources are offal and dairy products. Some green vegetables and wholemeal cereals are moderate sources and refined cereals contain only small amounts of riboflavin. But it has rarely been described in famine-afflicted populations and is not listed in standard works as a nutritional problem of emergency situations (de Ville de Goyet *et al.* 1978). One of the few descriptions is due to Smith and Woodruff (1951). They report that deficiency was observed in the Japanese military camps within 2 months of imprisonment, and in the civilian camps the deficiency (which was described as orogenital syndrome), was not described until 8 months of captivity. It may be that green vegetables grown by the prisoners made a sufficient contribution to the diet to account for the delay in onset of the symptoms.

However, riboflavin deficiency is not a spectacular deficiency disease, the major deficiency signs being dermatitis at the corners of the mouth and around the scrotum (Yagi 1978), and it would be easily overlooked. It may be for this reason that riboflavin deficiency has not been reported from any other famine situation. It is worthwhile investigating its possible existence further.

FOLIC ACID DEFICIENCY

Folic acid deficiency is probably the most common vitamin deficiency in the world (Hoffbrand 1978), but to judge from field reports apparently does not occur in famine. It is difficult to believe that this is so, since animal products are a primary source of the vitamin, and would generally not be consumed.

The most evident, but not pathognomonic, sign of folic acid deficiency is a macrocytic anaemia (Hoffbrand 1978). In the Minnesota experiment (Keys *et al.* 1950), and in the Warsaw Ghetto (Szjenman 1946), and even in the far worse conditions of the concentration camps (Collis 1945; Mollison 1946; Adelsberger 1946) the widespread anaemias, though of unknown aetiology (Keys *et al.* 1950), were essentially normochromic and normocytic with some cases being marginally macrocytic. In the Japanese internment camps, pregnant women had a tendency to macrocytic anaemia (Smith and Woodruff 1951), as they have normally. Macrocytosis was reported in the Dutch POWs in Japanese camps. Based on this evidence alone it might be concluded that if folic acid deficiency is present in famine, it is not generally widespread or severe.

However the macrocytic anaemia of folic acid deficiency is due to aberrations in the synthesis of nucleic acids in the deficient state (Passmore and Eastwood 1986), and it is possible that this is masked in famine. For example, if, in starvation, nucleic acid synthesis is depressed, generally or in the bone marrow, so that this sign of folic acid deficiency is masked, and a severe macrocytic anaemia does not develop.

Some clinical signs compatible with folic acid deficiency have been described in famine victims. For example glossitis was seen in concentration camp victims. In the Warsaw Ghetto, Stein and Fenigstein (1946) observed considerable haemosiderin deposition in soft tissues on *post-mortem*, something again characteristic, but not pathognomonic, of folate deficiency. One characteristic of severe famines consistently noted in light-skinned peoples has been patches of dark pigmentation of the skin, distinct from pellagrous lesions (see Hunger Disease). It seems possible that these are due to melanin deposition which occurs secondary to a folic acid deficiency (Hoffbrand 1978). These are only speculations, but the possibility that folic acid deficiency occurs in famine deserves further study.

PANTOTHENIC ACID DEFICIENCY

Pantothenic acid is at the other end of the scale from folic acid, being so widely distributed that deficiency would seem theoretically unlikely. However, in both the Japanese camps (Smith and Woodruff 1951) and the German concentration camps (Collis 1945; Mollison 1946) there was a strange disease aptly named 'burning feet syndrome' which seems from the description of symptoms to have been a pantothenic acid deficiency (Nelson 1978). One interesting aspect of the disease in the Japanese camps was that the patients affected rarely had beriberi (Smith and Woodruff 1951). Whether the disease was due to simple deficiency or the effect of an interfering anti-metabolite is not known.

MINERAL METABOLISM AND STARVATION

There are various consequences of famine for mineral metabolism. The possibility of demineralization of the bone secondary to a phosphataemia resultant on metabolic acidosis has already been mentioned. Keys *et al.* (1950) reviewed the literature and concluded the prevalence of such a hunger osteomalacia was very variable in different famines, and I think that is still a statement of the situation, though I repeat that little attention is currently paid to it.

Salt (sodium chloride) intake may be low if salt is a scarce, expensive commodity but direct deficiencies in sodium or chloride have not been recorded (though it may occur secondarily to diarrhoea when it is almost invariably fatal (Robson 1972)). Excessive salt intake may also be a problem particularly in populations consuming large amounts of soups (Widdowson 1985). Some observers are convinced that famine oedema only develops when salt is included in the diet (see McCance 1951).

Potassium deficiency is quite unlikely as there is likely to be an increased intake consequent upon the extent to which vegetable products dominate the diet.

Iron deficiency seems likely to be a major problem both because of a low iron intake and because the high plant-fibre intake will reduce iron absorption. As already noted, anaemia is common in famine but its dietary cause is obscure. Where iron deficiency anaemia is normally present to some degree, it will continue to be a problem in famine-affected populations.

There is no information on trace element deficiencies in famine.

HUNGER DISEASE

The above list of the principal ancillary diseases that accompany famine does not of course fully describe the diseases of famines. But it does deal with many of the symptoms described by classic authors under the eloquent, if general, description of hunger disease. Having dealt with these a hunger disease still remains: the collection of signs and symptoms that can reasonably be attributed to severe energy deficiency. These have been described in a variety of reviews, of which Keys *et al.* (1950) is the most extensive and Widdowson's (1985) is the latest.

The clearest and most unequivocal sign of inadequate food intake is loss of weight and in particular of body fat (Fig. 2.3), but also, to differing extents, of all the organs of the body, except apparently the brain (Table 2.4). Fliederbaum *et al.* (1946) and Widdowson (1985) have suggested that distinct stages of starvation can usefully be distinguished in adults on the basis of weight loss. Widdowson's scheme includes moderate starvation, when up to 25 per cent of the normal body weight is lost, severe starvation (<50 per cent body weight lost), and a terminal stage when weight loss exceeds this and the condition has so deteriorated that death often occurs despite treatment. Though these living dead, called in the concentration camps '*Musselmen*',* can be distinguished from those in the earlier stages, it is only sometimes practicable to generally draw other distinctions on the basis of weight loss, although a general progression in severity exists. The following description culled from various sources, especially Lipscombe (1945); Apfelbaum (1946); Leyton (1946); Fliederbaum *et al.* (1946); Apfelbaum-Kowalski *et al.* (1946); Keys *et al.* (1950); Winick (1979); and Widdowson (1985) outlines the symptoms observed.

The skin is pale, probably because of a reduced peripheral blood flow, and subjects tend to complain of the cold. The skin is dry, scaly, inelastic, and very thin with a tendency to ulcerate. Folliculosis is generally present (but see vitamin C, vitamin A, and EFA deficiency). Skeletal muscles are slack, movements are slow and clumsy. Even before there is a detectable loss of cardiac tissue, the pulse rate and the cardiac volume are reduced when measured at rest but disproportionately increased by quite mild exercise (Glaser 1951). As weight loss proceeds, the resting pulse is further reduced and blood-pressure falls. As weight is lost there is a progressive fall in the basal metabolic rate (BMR), and as weight loss proceeds a fall in BMR per unit weight. The alimentary canal is thinned, but gut function is generally normal in mild starvation. In severely starved subjects, however, gut function is disturbed so that feeding may be associated with severe diarrhoea, and food passes through the gut undigested (Mollison 1945). Urine volume is often reported as increased, possibly because reports come from situations where large volumes of soup have been consumed. Nocturia and polyuria are a frequent complaint, but no convincing kidney abnormalities have been demonstrated, even though kidney mass is lost. In moderate starvation, extra-cellular fluid (ECF) volume is somewhat increased, and some individuals, especially adults aged over 40 years, get hunger oedema. As greater amounts of weight are

* The term is Yiddish literally meaning a Moslem, and by extrapolation someone residing in a distant land. In this case the distant land in which they were living was the land of the dead (Keneally 1983).

lost the prevalence of hunger oedema rises, and all age groups become affected. The skeleton is radiologically normal, but there is some reduction in bone marrow which is replaced by water, which, being more radiologically opaque, makes the skeleton radiologically dense, and may mask demineralization. A dirty brown darkening of the skin has been frequently reported in light-skinned famine victims (but see folic acid deficiency). This discolouration may particularly occur in areas of the skin most traumatized. It is not due to exposure to the sun, although there is also an increased sensitivity of the skin to sunlight. The cell-mediated immune response is slightly reduced, with a consequently increased vulnerability to infections. In adolescents puberty is delayed, and may not occur. When it does it has unusual features. In boys the voice may not break and in both sexes there is a curious tendency in the more severely starved for a luxuriant growth of hair on the face and in the genital regions, and excessive amounts of a fine languo hair on the face, trunk, and limbs have been described (Keys *et al.* 1950; Widdowson 1985; Woodham-Smith 1962). In African populations the only sign of this observed has been the growth of long silky eyebrows (Seaman, personal communication). Severely starved adults from the age of 30 onwards may lose hair even in the armpits and genital regions (Fliederbaum *et al.* 1946).

Fertility is reduced, both by amenorrhoea and a reduced sperm count, and such children as are born tend to be underweight and of reduced viability. Lactation is not, however, necessarily impaired.

Body temperature is reduced. Keys's moderately starved volunteers had a mean fall in oral temperature of 0·25°C, while in the Warsaw Ghetto severely starved subjects had oral temperatures of 1·3–1·4°C below normal (Fliederbaum *et al.* 1946). It has also been noted that severely starved subjects with typhus do not show fever (Fliederbaum *et al.* 1946).

Creatinuria is reported in severe starvation, and after great muscular effort in moderate starvation, but otherwise urine chemistry is normal. Pupils are narrow and round and often react sluggishly to the light, there are senile changes in the lens, and intraocular pressure is reduced. There is severe anaemia (the RBC count being reduced by 50 per cent in severe starvation), but little evidence of increased red cell production (Szejenman 1946). The anaemia is at least partly due to increased red cell breakdown: *post-mortem* studies of deaths from starvation found haemosiderin deposits in many organs (Stein *et al.* 1946). There may or may not be an increased erythrocyte sedimentation rate.

Many reflexes are abnormal and psychological changes are remarked upon even in moderate malnutrition. There is a loss of libido, the patient is lethargic, with an overall impression of dullness and apathy. 'The patients have a characteristic masklike appearance. Their faces are expressionless and pale' (cited by Fliederbaum *et al.* 1946). Memory may be impaired. In the severe state there may be a complete breakdown in normal personality and in family relationships. The Musselman was often observed as being sometimes too apathetic even to collect his food (Fliederbaum *et al.* 1946).

Besides all these positive symptoms some diseases are notably absent. In the Warsaw Ghetto the physicians particularly remark upon the absence of acne, stomach ulcers, gastritis and ileitis, and liver or gall bladder ailments, with pre-existing sufferers from the latter two often noting an improvement in their condition (Apfelbaum 1946; Winick 1979; Fliederbaum *et al.* 1946). Eczema and psoriasis also improve (Netherlands Red Cross Feeding Team 1948). Murray *et al.* (1976) have noticed the absence of some infections with an increased susceptibility appearing on refeeding.

The following passage from Fliederbaum *et al.* (1946) describes the progression better than I can:

... Boys and girls change from blooming roses into withered old people. One of the patients said 'Our strength is vanishing like a melting wax candle.' Active, busy energetic people are changed into apathetic, sleepy beings, always in bed, hardly able to get up to eat or to go to the toilet. Passage from life to death is slow and gradual, like death from physiological old age. There is nothing violent, no dyspnea, no pain, no obvious changes in breathing or circulation. Vital functions subside simultaneously. Pulse rate and respiratory rate get slower and it becomes more and more difficult to reach the patient's awareness, until life is gone. People fall asleep in bed or in the street and are dead in the morning. They die during physical effort, such as searching for food, and sometimes even with a piece of food in their hands.

ENERGY BALANCE IN STARVATION

Weight loss in starvation is due to loss of both body fat and protein catabolized as fuels, and the concomitant loss of body water as protein is lost (Keys *et al.* 1950; MacDonald 1984; Widdowson 1985). Carbohydrate stores take only a few days to deplete and are unimportant in this context.

Much attention has been focused on two particular aspects of energy balance in starvation: whether there is an adaptive reduction in energy expenditure, and the extent to which adaptive mechanisms exist which specifically reduce protein losses. Neither of these questions has yet been fully resolved, not least because a central difficulty persists in agreeing upon the nature of adaptation. Adaptation is usually seen as something beneficial to the organism, and controversy exists about the extent to which changes that undoubtedly do occur in starvation are beneficial, or even harmful (Waterlow 1985). For example, one of the factors that reduce the energy expenditure of a starving man is a reduction in voluntary activity. Whether this should be seen as adaptive (i.e. beneficial) or pathogenic, evidence of abnormal behaviour, is a moot point.

There is no doubt that energy expenditure is reduced in starvation. Quantification of the extent of that observed change, however, is less unequivocal, discussion often tending to founder upon the rock of allometry, the impact of body size on physiological function. Size is a dominant factor in determining energy and protein needs and, all other things being equal, small animals need less food than large ones. Therefore a starved subject expends, and consequently needs, less energy as a result of weight loss. Disagreement persists as to whether expenditure is reduced in addition to this.

This can best be discussed by considering Basal Metabolic Rate (BMR), which is a measure of the obligatory rate of energy loss. BMR is defined as the metabolic rate of a subject who has not eaten for at least 12 hours and who is in a thermoneutral environment and free from stress, and has proved to be a very reproducible measurement in normal circumstances. In starvation, BMR declines as weight is lost, and BMR related to the usual indices of size (per unit body weight, per unit surface area, per $W^{0.75}$) falls also, but as Fig. 2.2 illustrates to different extents. Almost whatever

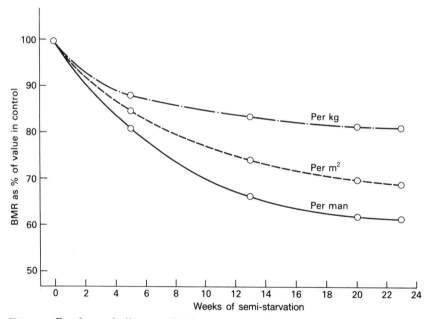

Fig. 2.2. Basal metabolic rate (BMR) (mean for 32 men) before and during 24 weeks of semi-starvation. All values are expressed as percentages of the pre-starvation (control) values for the oxygen uptake per man, per square metre of body surface, and per kg of body weight. Values per $W^{0.75}$ are intermediate between those per kg of body weight and per unit body surface area. Data from the Minnesota Experiment (Keys *et al.* 1950).

index is used, the extent of the fall depends upon the extent of the weight loss.

Data collected in the Warsaw Ghetto by Fliederbaum *et al.* (1946) give the most extreme range of values. These showed a 10 per cent reduction below normal in first degree cachexia (moderate starvation), in the second and third degrees of cachexia (severe starvation) the fall in BMR was 30–40 per cent below normal, and in two terminal patients, Musselmen, the BMR was 60 per cent below the standard. (The standard is not mentioned but the results were probably related to surface area calculated by Du Bois's method.)

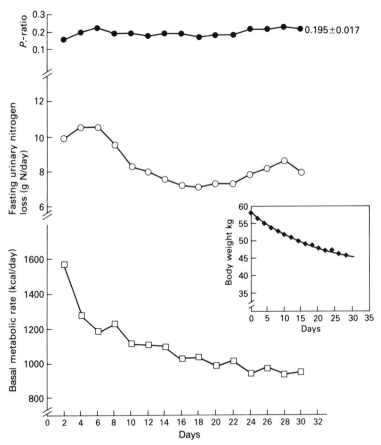

Fig. 2.3. The decline of body weight, basal metabolic rate, and protein catabolism (as judged by fasting urinary nitrogen excretion) in total starvation. However, the fraction of energy expenditure contributed by protein (the *P* ratio) remains constant. Data for subject Takagaroma, taken from Henry (1984).

Most investigators have studied subjects with smaller amounts of weight loss, and not all have managed to demonstrate for a population a relationship between the extent of weight loss and the reduction in BMR (Berkmann 1930; Beattie and Herbert 1947). Where longitudinal studies have been conducted BMR and body weight have declined together as illustrated in Fig. 2.3.

The decline in BMR does not exactly parallel that of body weight for two reasons. First there is a decline in the metabolic intensity of specific tissues with starvation, and secondly different organs (with different metabolic activities) are lost at different rates relative to each other and to overall weight loss. The first point was illustrated by Kleiber (1975) in some studies of experimental starvation in rats. He found that the *in vitro* oxygen consumption per unit weight of different organs from starving rats was considerably below normal: for diaphragm the decline was 31 per cent, for liver 59 per cent and for brain 53 per cent. Such marked reductions would probably be due in part to changes in the water content of the organs in starvation, in particular to a rise in extracellular fluid (ECF) and a decline in active tissue mass. It is not known whether comparable declines are observed in humans but it seems likely that they occur.

The other factor accounting for the change in BMR in starvation is the differential rate of tissue and organ loss illustrated in Table 2.4. Adipose tissue, which has a low metabolic rate, is lost, so that the mean metabolic rate of the residual tissue rises. This is counterbalanced in part by the fact that as weight is lost a higher fraction of body weight is made up by the

TABLE 2.4. Estimated loss of weight of various tissues and organs in starvation*

Organ	Normal weight (g)	% of normal weight lost
Skeletal muscle	28 000	40
Adipose tissue	9 750	> 90
Liver	1 750	55
Gut	2 000	60
Brain	1 350	< 5
Kidneys	150	25
Heart	280	40
Pancreas	100	40
Spleen	200	50
Blood–Plasma	3 200	− 10
–Erythrocytes	2 800	50

* Assumed: initial body weight 70 kg; weight loss in starvation of 40% and no visible oedema.

Sources: Author's estimates based upon *post-mortem* studies reviewed by Keys *et al.* (1950) and data presented by Stein and Fenigstein (1946).

skeleton which has an extremely low metabolic rate, and to some extent by the fact that, as Widdowson (1985) has suggested, water content of tissues of even non-oedematous subjects will increase somewhat, lowering tissue metabolic rates per unit weight.

The five metabolically most active organs – brain, liver, heart, kidney, and skeletal muscle – account for almost all (94 per cent) of the BMR of normally fed subjects. Simply by using the weight losses of these organs per unit body weight loss shown in Table 2.4, and allowing for a rise in the volume of the ECF, it is possible to make a good estimate of the BMR of a starved man without assuming changes in the metabolic intensity of these tissues, as Table 2.5 shows. Thus only a small (15–20 per cent) decline in the intensity of metabolic turnover need be assumed to make prediction agree quite well with observation. This is all purely speculative, but is worth investigating, not least for the possible implications of changes in the extent to which different organs come to dominate the energy economy of the body (Table 2.5).

TABLE 2.5. Basal metabolic rate in starvation and its possible partition between major organs

	Normal subject	After 20% weight loss	After 40% weight loss
Observed BMR* (% normal)	100	65	50
Predicted BMR†	94	74	63
Observed BMR % predicted	106	88	79
Estimated percentage BMR due to:†			
Skeletal muscle	28	25	25
Liver	26	21	17
Brain	23	27	32
Heart	10	9	8
Kidney	7	7	7

* Author's estimate from data given by Keys et al. (1950); Benedict (1915); and Fliederbaum et al. (1946).

† Predicted BMR is the sum of the 5 organ BMRs. Normal partition of BMR between organs from Henry (1984). Partition of BMR in starved subjects calculated assuming organ weight loss at 40% body weight loss as shown in Table 2.4; organ weight loss at 20% body weight loss derived by linear interpolation. A 10% increase in ECF assumed at 20% body weight loss (Keys et al. 1950); a 15% increase in ECF is assumed at 40% body weight loss. No decrease in metabolic intensity of organs, apart from that due to changes in ECF, is assumed.

PROTEIN CONSERVATION IN STARVATION

Little attention has been paid so far to the problem of protein deficiency in famine. This is because I do not believe there is one. High protein foods have been major, but unnecessary, items of relief aid in famines. The usual justification of this, that starving people need protein to replace lean tissues lost, neglects the fact that they need energy too, and the energy costs of growth are so high that only a small fraction of the calories need be supplied as protein. By contrast, the levels of protein in the relief foods have been more likely to kill than to cure. In one German relief centre in Ethiopia in 1974, a food was fed which provided 90 per cent of its calories as animal protein; as Seaman et al. (1974) noted, it was probably defatted offal.

There is an obvious commercial drive behind providing such foods, offal being something which had little commercial value in donor countries except as an ingredient of petfoods, and food aid made from it was therefore charity on the cheap. Similar motives probably account for the US government-funded initiative to produce high protein food from fish waste (Pariser et al. 1978) and the fact that dried skimmed milk is still a much overstressed aid item. But it was not commerce alone. There was also a mystique attached to protein which nutritionists did much to foster and sustain, until called into doubt in a succession of papers during the 1970s of which McLaren's (1974) 'The Protein Fiasco' and 'The Protein Gap' (Waterlow & Payne 1976) are the best known. For the last decade it has been accepted that few diets, even high cereal ones, provide insufficient protein, and that protein deficiency never occurs.* The recent FAO/WHO expert committee (FAO/WHO/UNU 1985) challenges this view, but I do not believe their evidence.

The interrelationship between energy and protein requirements can be seen in the parallelism between obligatory protein loss (measured by the nitrogen excretion of subjects on a protein-free diet, the obligatory nitrogen loss (ONL)), and the obligatory energy losses (i.e. BMR), ONL approximating 2–3 mg N/kcal BMR. There is also a relationship between energy intake and protein metabolism, though there are different views as to what occurs.

Henry (1984) and Henry et al. (1984, 1988, and unpublished data) have argued on the basis of comparative evidence, fasting studies in humans and animal experiments, that nitrogen excretion in total starvation (which they measure by the fasting urinary nitrogen loss, FUNL) is higher than ONL, although it is still related to BMR and ONL. According to these

---

* An adult maintaining weight requires about 8 per cent of the energy as mixed proteins, and a fast-growing child about 10 per cent. If the protein is very good quality these figures would be reduced to 4 per cent and 6 per cent. Most diets in the world provide 10–15 per cent of the energy as protein, cereals alone provide 8–11 per cent.

authors, FUNL is 1·5 times ONL, i.e. 6 mg N/kcal BMR. On the assumption that all this nitrogen comes from protein, this indicates that 15 per cent energy expenditure in fasting comes from the catabolism of protein.

Henry *et al.* have reanalysed classic studies of total starvation in volunteer humans and shown that, after the first few days of fasting, the FUNL declines in parallel to the fall in BMR (Fig. 2.3). Thus, the FUNL/BMR ratio (and hence the importance of protein as a fuel which they measure by the P ratio) is for a given subject, constant, at least until the weight loss is quite severe. At this point there is a rise in the relative importance of protein as a fuel, something described in the classic literature as the pre-mortal rise in protein metabolism.

These results stand in contradiction to the experimental studies of Felig, Cahill, and their colleagues on obese patients undergoing prolonged fasting for weight loss (Felig *et al.* 1968; Cahill 1970, 1978). These authors suggest that as fasting was prolonged, fat became a progressively more important fuel, so that levels of nitrogen catabolism fell far below even the ONL value. Felig's theory is that these low levels of protein catabolism are possible because the brain uses ketone bodies, rather than glucose, as fuel, thus reducing the need for gluconeogenesis from amino acids, which would otherwise dictate the rate of protein catabolism. In this view, the primary adaptation to starvation is a reduction in protein catabolism, a view whose logical attractions are illustrated by Widdowson's (1985) analysis:

In fact a man has in his body plenty of energy to keep him going for more than 70 days [of total starvation]. Let us suppose he weighs 70 kg and has 15 per cent of fat in his body – that is 10.5 kg of fat. If he oxidised this at a regular rate over 70 days 150 g fat a day would be broken down and this would provide him with 1350 kcal a day. It is not because he runs out of energy that a starving man dies. I am fairly sure that the Lord Mayor of Cork [who died in the 1920s after a hunger strike lasting 70 days] and his more recent counterparts still had fat in their bodies when they died. Rats that are starved to death certainly do. It is the catabolism of protein that goes on all the time that kills.

However, Henry *et al.* (1984) have pointed out that Felig's results are only based upon therapeutic fasting of grossly obese patients and may not therefore apply to the starving famine victim whose body fat is much lower. Their view is supported by their observation that the FUNL decreases with increasing body fat in humans and experimental animals. Equally it is possible that the subjects in the experiments upon which Henry *et al.* base their deductions had elevated nitrogen excretions which were elevated by, for example, acidosis.

It should be stressed that Henry *et al.* are not arguing with Felig's basic contention, that protein is conserved in starvation by a reduction in net catabolism. They are merely suggesting that the reduction is in direct proportion to the fall in energy turnover. Felig did not measure energy expenditure but the thrust of his case is that protein needs in starvation are reduced by even more than this because fat becomes a primary fuel. The results from both groups agree that there is no justification for feeding high protein foods in starvation.

ENVIRONMENTAL FACTORS AND FOOD NEEDS

The fact that famine-affected populations often have reduced amounts of clothing and limited shelter can be very important because of its impact on energy requirements (Rivers and Brown 1979; Seaman 1984).

Most of the energy humans require is used for internal work in the body, and so eventually appears as heat. In effect animals, human amongst them, burn food producing heat. That heat is lost from the body surface by radiation to cooler parts of the environment, by conduction to any cooler objects with which the body is in contact, and by the evaporation of water from the skin in sweating and in insensible perspiration. Humans are homeotherms: they try to maintain their temperature at a constant level of about 37°C, by regulating heat loss by physiological or behavioural processes, or, if necessary, by increasing the rate at which food is metabolized. It is the latter that might be particularly important in famine.

The temperature at which energy requirements begin to rise to meet the demands for 'cold-induced thermogenesis' is called the Critical Temperature (CT). For a western 'standard male' unclothed it is about 28°C in still air (Mount 1979). In a lightly clothed subject the CT is lower: about 25°C in still air. (The effect of wind is to increase heat loss, and the CT is increased with wind speed. A wind chill factor can be expressed by calculating an apparent temperature from the combination of true temperature and wind speed, in the manner described for example by Mount (1979). For simplicity all calculations presented here assume still air.)

Below CT, the rate of increase of heat production with fall in temperature will principally depend upon the deep body temperature (core temperature), the thermal conductivity of tissues, something influenced by fatness and blood flow, and adaptive responses such as huddling up (which reduces surface area). Experiments in the UK have shown that lightly clothed, normally fed adult subjects at 22°C have a metabolic rate 7 per cent higher than that at 28°C (Dauncey et al. 1981). Approximate calculations based on this figure suggest that the metabolic rate rises by about 30 kcal/°C below the CT over this range. The rate of increase is probably lower in these marginal subcritical temperatures due to the mitigating effects of adaptive responses at these marginally subcritical temperatures. I would expect faster rates of increase in metabolic rate at lower temperatures. Higher rates of increase would be found in famine, where subjects would have less body fat. At an environmental temperature below 20°C I estimate heat production in lean subjects would rise to a maximum of about 3–5 times BMR. Thus at 15°C the total energy expenditure might be more than 750 kcal/day higher than that at thermo-neutrality unless the subject is able to reduce heat loss by, for example, putting on extra clothing (Rivers and Brown 1979). At lower temperatures, despite elevated metabolic rates, hypothermia, an ultimately fatal reduction in core temperature would occur.

This thermoregulatory heat requirement need not come from food, it can be provided by direct radiant energy, sitting in the sun or by a fire. But in the absence of these possibilities, it must come from increasing the metabolic rate. For a well-fed Western subject, 750 kcal/day is equal to an increase in energy expenditure of about 20 per cent, but for an underfed or fasting subject it is about a 50 per cent increase. Unless the food intake is increased by this proportion starvation is exacerbated. Experiments conducted on young rats illustrate the potential importance of the effect. As Fig. 2.4 shows, rats, fasted at 20°C, died after 11 days of

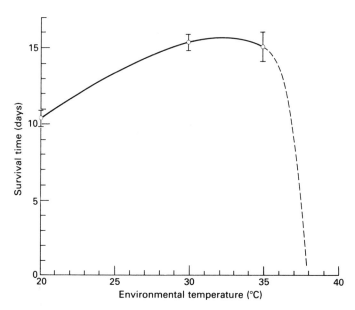

Fig. 2.4. The survival time of 100-day-old rats totally fasted at different environmental temperatures. Dotted line is based on data from separate studies of heat stress. Modified from Kleiber.

total starvation, those fasted at 30°C and 35°C (CT for a rat would be about 30°C) lasted 17 and 16 days, a non-significant difference. The more rapid demise of the cold-stressed rats was due to their higher metabolic rate which was 1·4–2·0 times that of the thermoneutral controls (Fig. 2.5). Kleiber also shows that the cold-stressed animals had lower respiratory quotients, indicating greater reliance on body fat as fuel, i.e. a lower P-ratio.

The CT depends upon the ratio of the rate of heat production to dissipation. Heat is lost through body surface and the smaller someone is the more body surface per unit weight they have, so that the higher will be their CT. Thus small people may be more sensitive to the cold than big ones. A new-born child, for example, has a CT of 33°C, and is, therefore, extremely vulnerable to hypothermia. At any body size the lower the heat production, the higher will be the CT. Clearly, famine victims in general are likely to have elevated CTs because of their lower metabolic rate.

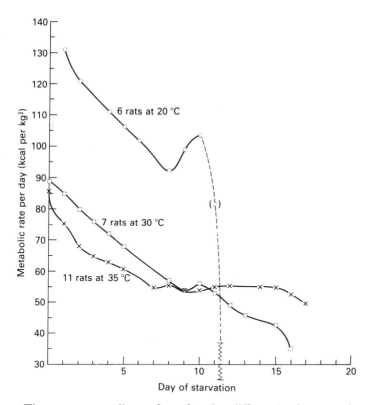

Fig. 2.5. The energy expenditure of rats fasted at different environmental temperatures. Results are expressed relative to W(0·75) to correct for differences in body weight. From Kleiber.

Additionally, the loss of subcutaneous body fat reduces the insulating properties of the skin and increases heat loss across the skin at any temperature. On these grounds famine victims can be expected to have a higher CT.

Although I know of no direct measurements on this, in famine, the probable lack of food, clothing, shelter, and fuel combine together to make it very likely that the famine victim will be cold stressed. The problem is not one of cold areas alone. Measurements I made in northern Ethiopia, some years before the 1973 famine, suggested that a lightly clothed underfed adult would be cold stressed for more than 12 hours a day, unless able to shelter or wrap up well, or bask in the sun (Rivers and Brown 1979).

A good indication of whether a person is below CT is whether they feel uncomfortable from the cold. If they are shivering visibly they are quite significantly cold stressed. A frequent report from famines is the extent to which famine victims complain of the cold, although the severely malnourished may be too apathetic to complain.

Famine victims may have their vulnerability to cold stress reduced by two factors. The first is the fact that oral temperature is lower (see Hunger Disease). If this reflects a lower body core temperature and not just differences in heat flow to the mouth, the CT would be lowered. The second ameliorating factor is that the malnourished adult has reduced peripheral blood flow, particularly to the limbs, this is why they are called pale. This will tend to reduce heat loss across the body surface.

Cold-stressing the famine victim may not necessarily increase heat production. Infants and the elderly if cold stressed tend instead to reduce body temperature, and this may happen in famine victims. Hypothermia does lower the CT but is not an adaptive response. It cannot be sustained without adverse effects, indeed ultimately it is fatal. Seaman (1984) has suggested that it is connected with the tendency for mortality after natural disasters to be highest amongst the old and the young.

In dealing with famine situations it is important not to minimize the extent to which cold stress contributes to energy needs for all age groups, not least because in the last analysis reducing heat loss is on all accounts less expensive and more manageable than providing food as fuel to provide the heat. A blanket which reduced the heat loss by cold stress in an environment of 15°C could save food equivalent to 1–2 kg of grain per adult per week, in other words equal to its own weight. The implications of this for the logistics of freighting supplies do not need enlarging upon.

DEMOGRAPHY OF FAMINE

There is no doubt that famine kills selectively, with young children amongst its most publicized victims. According to one report from the recent East African famines, only about 6 per cent of the population reaching feeding centres were young children; pre-famine, the population would have had about 15 per cent in this age group. The implied death-rates are not unusual. However, deaths in a famine are the outcome of two separate factors, physiological vulnerability and social protection. Young children are more vulnerable to nutritional deprivation, they have smaller reserves of nutrients and energy, and appear less able than an adult to adapt to shortage. A semi-starved adult loses weight, but may eventually re-establish energetic equilibrium at a lower level of expenditure, as any would-be slimmer will testify. Although an underfed child only loses weight if severely deprived, and in more mild shortages merely grows more slowly, neither state is stable. Rather the child seems to become more vulnerable to all manner of diseases as it falls further and further behind its expected size. The consequence is that mortality in children rises with severity of PEM.*

Old people, too, seem to be physiologically vulnerable to famine. As already noted, it is the old who show signs of famine osteomalacia and the old who seem particularly vulnerable to hypothermia.

Western cultural prejudice would have us believe that women are more vulnerable to famine than men. However, there is little physiological justification for this view. Indeed as I have argued elsewhere (Rivers 1982), females ought to be less vulnerable to deprivation, having smaller needs for energy and most nutrients because they are smaller than men, have a lower metabolic rate and a higher body fat. Laboratory experiments certainly often show that male animals are more vulnerable to deprivation than females (see, for example, Rivers 1982; Rivers and Crawford 1974).

Physiological vulnerability is, however, modified by social factors which can totally reverse these expected effects. Although children are more physiologically vulnerable, increased death rates among children are not the invariable consequences of famine. They are only famine traits in so far as the family is not able to choose to sustain the child at the expense of the adults. This has sometimes been the case; there is for example no increased mortality associated with the 'Hunger Winter', the Dutch Famine of 1944–5 (Stein et al. 1975). Such differences undoubtedly reflect in part value systems of different cultures, but also, of course, the severity of the famine. In severe deprivation it is implausible for the adults to sustain the children at their own expense simply because not enough food exists to do this. This was observed in the Warsaw Ghetto where even given the child-centred orientation of the population, infant and child mortality rose during the imprisonment to levels higher than those observed in develop-

* Risk does not rise linearly with shortfall in growth. Indeed it may be that there is a threshold effect, with risk only rising beyond a certain degree of deficit (see Kielman et al. 1978; Chen et al. 1980; Pacey and Payne 1985 for discussion).

ing nations. And of course in severe starvation there is a breakdown of personality such that, for example, parents will steal from their children to obtain food.

Moreover despite their possibly lower physiological vulnerability, there is evidence that in famines, as, in other types of disasters, there is an increased morbidity and mortality amongst adult women, and amongst female children, presumably reflecting male discrimination against them (see Rivers 1982). Indeed, even where there is no sex difference in death rates or nutritional status, male discrimination can still be inferred from the fact that females should be less vulnerable (Rivers 1982).

There is, unfortunately, little information in this important area, and it would be most useful if more attention was paid to it, in particular in future famine situations.

What nutritionists mean by the nutritional status of communities is in a rather ill-defined way the extent to which the community is suffering from food shortages and malnutrition. Nutritional status used in this way is a particularly Boojumish snark, particularly difficult to quantify. It is not the average of individual nutritional status measurements. Attempts to do this quickly reveal bizarre value-judgements as necessary: how many cases of mild malnutrition were equivalent to a case of severe malnutrition, how far should adequate nutritional status in adults be offset by PEM in young children? It is really the perception of the investigator about how bad things are and how far the situation might change. The impressionistic investigations of a skilled operator can result in a description which, however tenuous, is none the less valuable. But it is a description which depends for its validity on the analytical skills of the investigator, and it is therefore one to which statistical measures of certainty cannot be attached.

Attempts to assess the nutritional status of communities in famine are difficult, because famines by their nature impose demands of speed in gathering information and for extensive surveys of a wide area. A shortage of skilled manpower has meant that, in the last decade, the notional assessment of the nutritional status of the communities has often meant little more than anthropometric surveys, that is, measurement of how big the children are. The stress given to this varies between agencies. Those who employ genuine experts, skilled in field-work, tend to lay great stress on the need to define famine in terms of its socio-economic impact, though anthropometry still figures as a major tool. Others rely principally on anthropometry. There are simply not enough genuine experts, though if the plethora of agencies involved in assessment were to co-ordinate their activities, the available manpower might be sufficient. But it is also my impression that a powerful lobby exists, especially amongst American epidemiologists and nutritionists, and in sections of the UN agencies, that would prefer to place primary stress in famine assessment upon the anthropometric assessment. Some experts indeed argue that famines should be assessed solely by the extent and severity of PEM in children, that is, on the characteristics of their growth and growth failure.

I think this has come about for a nexus of reasons, which are quite understandable, but nevertheless wrong.

The use of growth as a measure of nutritional status in children has a long, if chequered, history. Its use in developing countries can be traced back particularly to Professor Gomez's proposal that PEM in children be graded by the simple procedure of measuring how big they are relative to what might be expected. Because children get bigger as they grow older their expected size rather obviously depends on how old the child is, so that assessment requires two measures: body size (weight in Gomez's scheme) and age, the calculated parameter, weight for age (WfA), being weight relative to that found at that age in a reference population (Gomez et al. 1955).

But age is a problem: because normal children grow quite fast, small errors in estimating age can lead to quite large errors in apparent nutritional status. In the 1967 Biafran famine, where anthropometric assessment was used widely for the first time to assess the nutritional status in a famine, the field workers found they were often dealing with people, particularly orphans, amongst whom only the vaguest idea of age existed. In such circumstances it is no use estimating ages 'by eye', for estimates of the age of young children have as their basis preconceptions about the relationship of size and age, which are invalidated by malnutrition. Apparent nutritional status was, therefore, a product of errors in the assessment of age as well as of growth failure.

The American Friends' Service Council (FSC) popularized the assessment of nutritional status in these circumstances by relating attained growth not to age, but to height (Arnhold 1969). The argument used was simple, but elegant. In famine, PEM is more severe than that found in underdevelopment, so the child does not simply grow slower, it actually loses weight. But it was argued height cannot be lost; no matter how severe the deprivation, the rate of height gain can only be reduced to zero. The famine child loses weight but not height and therefore appears skinny, that is, underweight relative to height. The FSC made a further contribution by assessing weight indirectly, by measuring arm circumference as a proxy and introduced the use of the upper arm circumference:height ratio (the QUAC stick method) as a method of assessing which children were in need of nutritional intervention, but the principle still applies.

In 1972, Professor Waterlow put this *ad hoc* methodology on a much firmer footing when he proposed that in general, malnourished children could be divided into two groups by comparing body weight with height, and height with age. Malnourished children have low attained growth for their age; Waterlow suggested:

[1] height should be used as the indicator of preferred measure of growth, and children of inadequate attained height for their age be called STUNTED;

[2] that it was worthwhile to discriminate children who had low weight for their height, regardless of their degree of stunting. These children Waterlow called WASTED.

Wasted children are a minority of the malnourished in a developing country, and the axiom became accepted that the normal malnourished child, though small, is perfectly proportioned, that is, that mild chronic malnutrition affects weight and height gain in proportion. The wasted child was seen as the result of acute severe malnutrition, where weight loss occurred.

Waterlow's theory and methodology has been accepted and enlarged upon by various expert committees (see, e.g. WHO 1983) and it has become widely used by operational agencies and field workers in famines as in chronic situations. Famine epidemiologists, like almost everyone else, generally accept now that anthropometric assessment can be used to discriminate acute and chronic forms of malnutrition, and, therefore, weight for height (WfH) has become extensively used in the assessment of the extent of famine and the evaluation of the efficacy of famine relief.

During this process the anthropometric assessment has undergone a subtle transformation. As pioneered by the FSC in Biafra, WfH was used as a method of deciding upon priorities for feeding, by workers who had insufficient supplies for the masses who swamped the feeding centres. Since risk of dying rises quite steeply below a certain degree of wasting it is a good tool in this regard (Chen et al. 1980; Pacey and Payne 1985).

Subsequently, however, it has become used as a preferred diagnostic tool for measuring the *extent* of the problem. Such recent agency reports from the Ethiopian crises that I have seen, and, I am sure, the vast majority I have not, use WfH in children as their primary measure of the severity of famine.

I think that the enthusiasm may have been premature, and, since so much data has now been gathered, it may be possible to use this to re-evaluate the procedure. I cite here two examples to show why this may be necessary. The first, in Table 2.6, comes from Sudan in 1984–5, and has been kindly made available to me by Dr Behrens. In two different regions at two different times within the famine area, the prevalence of wasting in children varied from a low of 9 per cent to a high of 23 per cent. Such variability could of course indicate that a threefold variation existed in the severity with which different communities were affected by famine.

TABLE 2.6. Anthropometric results and market prices in surveys in 2 regions of Sudan, 1985

	South Kordofan		North Kordofan	
	May/June	Sept./Oct.	May/June	Sept./Oct.
% children wasted*	9·0	10·6	23·2	11·9
Range	(3·3–18·4)	(0–20·0)	(8·3–28·4)	(0–35·0)
Number of villages	8	8	9	9
Grain price % normal	250	150	260	160
Animal price % normal	27	114	55	130

* Wasting was defined as <80% weight for height using NCHS standards. Values shown are means of village means obtained in survey.

Source: Data of Mohammed and Behrens, (Mohammed 1986), recalculated by this author.

But neither the rise in grain prices nor the fall in animal prices, both external indicators of the economic distress caused by famine, could be unequivocally related to the degree of wasting.

In Table 2.7, I present results we collected in the 1974 Ethiopian famine (Seaman *et al.* 1974, 1978; Holt *et al.* 1975; Rivers *et al.* 1976). Data are

TABLE 2.7. The relationship between wasting and socio-economic indicators of famine, Harerghe Province, Ethiopia, May/June, 1974

	Issa	North Ogaden	South Ogaden	Marginal cultivators	Highland
% children wasted (<80% wt for ht)*	10·4	25·2	12·6	14·2	19·0
Sample number	(126)	(305)	(292)	(134)	(151)
Adult absentees at time of survey†	16·9	34·2	12·9	6·6	3·8
Deaths in under-fives‡	30·6	24·2	29·0	14·2	9·2
Deaths in 5–14-year-olds§	5·5	4·3	5·9	3·5	1·5
% cattle dead	88·0	56·0	47·0	37·0	20·0
Severity of famine¶	very severe	severe	severe	mild	very mild

* Wasting defined as <80% weight for height using standards of Stuart and Stevenson.
† % reported absent for more than one month at the time of survey.
‡ As % births in last year.
§ As % mid-year populations.
¶ Author's estimate based on a range of social parameters.

Source: Data from Seaman *et al.* (1974).

shown for sample areas adversely affected by famine, essentially nomadic populations, and those predominantly agricultural areas which were far less severely affected by the famine.

It is clear that a difference in nutritional status did exist amongst the children not least because there was three times the death rate amongst children in the nomadic groups as amongst the predominantly agricultural groups, and we were able to demonstrate by considering socio-economic criteria that this division into famine-affected and famine-unaffected areas was correct. Yet on the basis of WfH alone there would appear to be no difference in the wellbeing of the two communities.

Such variability makes the use of wasting as a sole or preferred indicator of famine an extremely suspect activity, at least until the basis and implication for such variability is understood.

There are various possible explanations for such results, any or all of which might be appropriate in a given situation. Some are relatively obvious. For example, since the more severely wasted children are more likely to die, high mortality could paradoxically lead to an improvement in mean nutritional status. Another confounding variable is the community's attitudes towards children, that is, the extent to which they are preferentially protected.

The background, endemic, situation on which famine is superimposed may also affect the results though in a more complicated way. The assumed characteristic of PEM in children in developing countries is that more children are mildly affected and, therefore stunted, than are severely affected and also wasted. The use of wasting to assess famine assumes it is normally a rare phenomenon. But this is not so. Its extent varies in normal times and there appears to be a positive relationship between the extent of stunting in the community and the extent of wasting, as Martorell (1985) has shown. The relationship he detected (Fig. 2.6) is a curvilinear one with low levels of stunting (10 to 30 per cent) being associated with a broadly constant low level of wasting and thereafter wasting rising as stunting increased. Thus in non-famine situations, where stunting was highly prevalent, wasting was a much more important presentation, with about one child wasted for every four children stunted. Thus in so far as stunting measures nutritional status, it appears that the lower the average nutritional status of the community with no famine, the more wasting is found. It should be noted that some of the prevalences of wasting observed by Martorell in the normal underdevelopment were higher than those observed in famine.

Of course, since wasting carries with it an increased risk of death it could be argued that high prevalences of wasting demand intervention whatever their cause. Even if it were this simple, it ought to be clear that different kinds of interventions are needed for famine and underdevelopment and

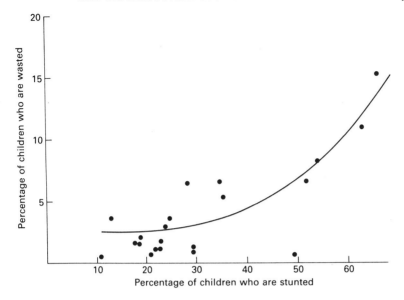

Fig. 2.6. The relationship between the prevalences of wasting (<80 per cent weight for height) and stunting (<90 per cent height for age) in underdeveloped countries, without famine. Data of Martorell (1985).

what is good for one could be disastrous for the other. It may seem that not even the most blinkered epidemiologist would suggest dumping food aid on any community in which wasting rose above a threshold value, regardless of the economic basis for the phenomenon, but the proposal has been seriously made by apparently responsible people. Even if the idea of using wasting as the sole indicator of intervention were to be entertained, decision-making is more complex than is often thought. The mortality risk associated with wasting depends on the extent of stunting: the risk being higher in children who are wasted and stunted than in wasted, non-stunted children. So low levels of wasting in a community where stunting is prevalent could indicate a higher prospective mortality than high levels of wasting in previously non-stunted children.

It could even be that we are caught in a semantic trap. Low WfH sounds important because we call it wasting, and thereby impose upon it a model of how we think it developed. But a child could develop a low WfH, without losing weight at all if what happened was that gain in weight stopped while gain in height continued. This does occur. Valverde (1975) found in Costa Rica that when the situation development projects were introduced in certain communities the prevalence of wasting rose because children started to gain height faster than weight. Dr Tony Nash of SCF found that this fall in WfH due to preferential height gain with apparently improving nutritional status also occurred in children in Quorem camp in Ethiopia in 1984 (Nash unpublished data).

These observations contradict the assumption that underlies anthropometric assessment, that, in all except severe deprivation, growth in height and weight are coupled. This axiom has never been proven and increasingly looks in error. For example, we have recently gathered evidence that in poor communities in the UK in the early years of this century, while height growth was much impaired, weight growth was more than adequate (Petty and Rivers 1986 and unpublished data). It seems that growth in height was restricted by wholly different factors from those which influence growth in weight. These may be nutritional. Some authors have speculated that stunting indicates protein deficiency, while wasting indicates energy deficiency (Malcolm 1970; Martorell 1985). Another view suggests stunting results specifically from inadequate zinc intakes (Golden 1985). I think it premature to support either of these theories, but the central and important point is that the axiom that under normal circumstances gains in height and weight are coupled can no longer be unquestioningly accepted, and without it, the basis for treating low WfH as wasting, i.e. loss in weight, disappears.

The other underpinning belief of anthropometric assessment of famine, that height cannot be lost, may also be untrue in famine. Measurements from the Japanese camps show that adults lost up to 2·5 cm in height due to severe undernutrition (Smith and Woodruff 1951). Ivanovsky (1923) reported that in the Russian famine adult men lost 3·8−6·9 cm in height, adult women 3·6−4·8 cm, in both cases mainly from the trunk. Height can be lost, as Smith and Woodruff point out, because of loss of soft tissues, such as cartilage of the intervertebral discs, or the heel pads of the skin. These data were collected on adults; whether comparable results apply to children is not known, but the possibility exists, and since it does the basis for using WfH for famine assessment is further weakened.

In the light of all such doubts it is particularly important to reconsider not only the appropriateness of different anthropometric indicators of famine, but the far-reaching effects of this particular approach in famine assessment. I have before me, as I write, a paper purporting to describe a nutrition survey in drought-affected areas of South Africa in July 1983. The authors conducted only WfH surveys of the most restricted kind, yet felt able to conclude:

No crisis could be demonstrated in those districts of Natal and Kwazulu which were selected for investigation on the strength of having experienced the worst drought (Kustner et al. 1984).

I find it worrying that such a relatively untried technique as assessing wasting has secured such widespread support. The belief that it measures loss of body substance, and therefore severe starvation, is untested in a way that would be unacceptable in, for example, a laboratory assay being used to diagnose disease in a Western hospital. Caution in using the technique is required not merely because, as I hope I have shown, the level of wasting varies inexplicably, but because we are, by advocating such an approach, continuing to ignore a basic and unexplored aspect of the nature of famine, the question of who has the right to decide on whether a famine exists.

Indeed, it is probably impossible to measure the importance of famine to its victims by its impact on health. In the 1973 Ethiopian famine it was the social impact of the famine that galvanized the world into action and, within Ethiopia, precipitated a chain of events that ultimately led to the overthrow of the Emperor, the institution of a Marxist regime and the revitalization of two rather ineffectual liberation movements into serious alternative governments over large parts of the country. Yet the death toll was only about 15 per cent above that which was the yearly death toll due to malnutrition and underdevelopment (Rivers *et al.* 1976). The real impact of the famine came from the refugees who fled the famine areas in rural Ethiopia and flooded down the roads towards the major cities. A minor local shortage became a major famine because the local resources were not adequate to provide for these people what the local culture made it reasonable for them to expect. This is the social dimension of the nutritional biology of famine.

CONCLUSIONS

This has been a long enough chapter without wasting time on repetition, under the guise of conclusions. Suffice to say that as a nutritionist who had come to believe that the study of famine had moved beyond my purview, preparing this review has revealed to me how much we have still to learn of its biology. I hope that this necessarily superficial review of the full range of those problematical areas might whet the appetite of some of its readers to provide some resolution of the doubts.

ACKNOWLEDGEMENTS

My thanks go to many individuals who contributed information, and with whom I have discussed at various times the ideas in this chapter. Though, doubtless, I got the information wrong and ignored their good advice, I thank in particular, Ms Judith Appelton, MBE, Ms Andrea Baron, Dr Ron Behrens, Ms Angela Berry, Dr Frances D'Souza, Rabbi Hugo Gryn, Dr Jeya Henry, Dr Tony Nash, Dr Celia Petty, Dr Philip Payne, Dr John

Seaman, and Dr Simon Strickland. Financial support from the Leverhulme Trust, the Butter Information Council, and the Wellcome Trust has made possible the work described here, and the compilation of this review.

LIST OF REFERENCES

Abedin, Z., Hussain, M. A., and Ahmed, K. (1976). Liver reserve of vitamin A from medico-legal cases in Bangladesh. *Bangladesh Medical Research Council Bulletin*, **2**, 43–51.

Ackroyd, W. R. (1930). Beriberi and other food deficiency diseases in Newfoundland and Labrador. *Journal of Hygiene* (Camb), **30**, 357–86.

Adelsberger, L. (1946). Medical observation in Auschwitz concentration camp. *Lancet*, **i**, 317–19.

Apfelbaum, E. (ed.) (1946). *Maladie de famine: recherches cliniques sur la famine executées dans le Ghetto de Varsovie en 1942*. American Joint Distribution Committee, Warsaw.

Apfelbaum-Kowalski, E., Pakszwer, R., Zarchi, J., Heller, A., and Askanas, Z. (1946). Recherches cliniques sur la pathologie du système circulatoire dans la cachexie de famine. In *Maladie de famine: recherches cliniques sur la famine executées dans le Ghetto de Varsovie en 1942*. (ed. E. Apfelbaum), pp. 189–225. American Joint Distribution Committee, Warsaw.

Arnhold, R. (1969). The QUAC stick: a field measure used by the Quaker service team in Nigeria. *Journal of Tropical Pediatrics*, 243–7.

Baker, E. M., Hodges, R. E., Hood, J., Saurberlich, H. E., March, S. C., and Canham, J. E. (1971). Metabolism of 14-C and 3-H labelled L-ascorbic acid in human scurvy. *American Journal of Clinical Nutrition*, **24**, 444–54.

Bartley, W., Krebs, H. A., and O'Brien, J. R. P. (eds) (1953). Vitamin C requirements of human adults, Medical Research Council Special Report Series, No. 280. HMSO, London.

Beattie, J. and Herbert, P. H. (1947). The estimation of the metabolic rate in the starvation state. *British Journal of Nutrition*, **1**, 185–91.

Beattie, J., Herbert, P. H., and Bell, D. J. (1948). Famine oedema. *British Journal of Nutrition*, **2**, 47–65.

Benedict, F. G. (1915). *A study of prolonged fasting*, publication No. 203. Carnegie Institute, Washington.

Berkmann, J. M. (1930). Anorexia nervosa, anorexia, inanition and low basal metabolic rate. *American Journal of Medical Science*, **180**, 411–24.

Cahill, G. F. (1970). Starvation in man. *New England Medical Journal*, **282**, 668–75.

Cahill, G. F. (1978). Physiology of acute starvation in man. *Ecol. Food Nutr.*, **6**, 221–30.

Carpenter, K. J. (1981). *Pellagra*. Hutchinson Ross, Stroudsberg, Pennsylvania.

Carpenter, K. J. (1986). *The History of Scurvy and Vitamin C*. Cambridge University Press, Cambridge.

Carpenter, K. J. and Lewin, W. J. (1985). A reexamination of the composition of diets associated with pellagra. *Journal of Nutrition*, **115**, 543–52.

Chen, L. C., Chowdhury, A. K. M., and Huffmann, S. L. (1980). Anthropometric assessment of energy-protein malnutrition and subsequent risk of mortality amongst preschool aged children. *American Journal of Clinical Nutrition*, **33**, 1836–45.

Collis, W. R. F. (1945). Belsen Camp: a preliminary report. *British Medical Journal*, **ii**, 814–16.

Curran, J. O. (1847). Observations on the scurvy as is has lately appeared throughout Ireland, and in several parts of Great Britain. *Dublin Quarterly Journal of Medical Science*, **4**, 83–134.

Dauncey, J. (1981). Influence of mild cold on 24 h energy expenditure, resting metabolism and diet induced thermogenesis *British Journal of Nutrition*, **45**, 257–67.

de Ville de Goyet, C., Seaman, J. A., and Geijer, U. (1978). *The management of nutritional emergencies in large populations*. WHO, Geneva.

De Wardener, H. E. and Lennox, B. (1947). Cerebral beriberi (Wernicke's encephalopathy): review of 52 cases in a Singapore prisoner-of-war hospital. Lancet, **i**, 11–17.

De Wijn, J. F. (1947). Waargenomen bij langdurige ondervoeding in krigsgevan-gen-kampan te Batavia en Fukuoko (Japan) gedurende de Japanes bezetting van Nederlandsch Oost-Indie (1942–45) in het bijzonder over Haematologische Bevindingen bij pellagra sine pellagra. Academisch Proefschrift, Universiteit te Verdedegen.

Desai (1988). This volume (Chapter 3).

Drummond, J. C. and Wilbraham, A. (1957). The Englishman's food: a history of five centuries of English diet (2nd edn). Jonathan Cape, London.

D'Souza, F. (1988). This volume (Chapter 1).

FAO/WHO (1973). Energy and protein requirements, Report of an ad hoc expert committee, WHO Technical Report Series, No. 522. WHO, Geneva.

FAO/WHO/UNU (1985). Energy and protein requirements, Report of an expert consultation, WHO Technical Report Series, No. 724. WHO, Geneva.

Fajgenblat, S. (1946). Les troubles oculaires dans la famine prolongée. In *Maladie de famine: recherches cliniques sur la famine executées dans le Ghetto de Varsovie en 1942* (ed. E. Apfelbaum), pp. 259–64. American Joint Distribution Committee, Warsaw.

Felig, P., Owen, O. E., Morgan, A. P., and Cahill, G. F. (1968). Utilisation of metabolic fuels in obese subjects. *American Journal of Clinical Nutrition*, **21**, 1429–33.

Fliederbaum, J., Heller, A., Zweibaum, K., Zarchi, J., Szejnfinkel, S., Goliborska, T., Elbinger, R., and Ferszt, F. (1946). Recherches cliniques et biochemiques sur les malades en famine. In *Maladie de famine: recherches cliniques sur la famine executées dans le Ghetto de Varsovie en 1942* (ed. E. Apfelbaum), pp. 79–172. American Joint Distribution Committee, Warsaw.

Fraser, D. R. (1981). Biochemical and clinical aspects of vitamin D deficiency. *British Medical Bulletin*, **37**, 37–42.

Glaser, E. M. (1951). Responses of the blood pressure and pulse rate to postural changes and exercise. In *Studies of undernutrition in Wuppertal, 1946–9*, Medical Research Council Special Report Series, No. 275, pp. 280–8. HMSO, London.

Glusman, M. (1947). The syndrome of 'burning feet' (nutritional melalgia) as a manifestation of nutritional deficiency. *American Journal of Medicine*, **3**, 211–33.

Golden, M. H. N. (1985). The consequences of protein deficiency in man and its relationship to the features of kwashiorkor. In *Nutritional adaptation in man* (eds K. L. Blaxter and J. C. Waterlow), pp. 169–85. John Libbey, London.

Goldsmith, G. A., Sarrett, H. P., Register, J. D., and Gibbens, J. (1952). Studies of niacin requirements of man. I: Experimental pellagra in subjects on corn diets low in niacin and tryptophan. *Journal of Clinical Investigation*, **31**, 533–42.

Gomez, F., Galvan, R. R., Cravioto, J., and Frenk, S. (1955). Malnutrition in infancy and childhood with special reference to kwashiorkor. *Advances in Pediatrics*, **7**, 131–51.

Hehir, P. (1922). Effects of chronic starvation during the siege of Kut. *British Medical Journal*, **1**, 865–8.

Henry, C. J. K. (1984). Protein: energy interrelationships and the regulation of body composition. Ph.D. Thesis. University of London.

Henry, C. J. K., Rivers, J. P. W., and Payne, P. R. (1984). Protein and energy metabolism in starvation reconsidered: is protein conservation really the crucial adaptation? *Nestlé Foundation Annual Reports*, pp. 71–9.

Henry, C. J. K., Rivers, J. P. W., and Payne, P. R. (1988) Protein and energy metabolism in starvation reconsidered. Hum. Nutr: Clin Nutr. *42C* (in the press).

Hess, A. F. (1920). Scurvy past and present, Nutrition Foundation Reprint Series, 1982, Academic Press, New York.

Hodges, R. E., Hood, J., Canham, J. E., and Sauberlich, H. E. (1971). Clinical manifestations of ascorbic acid deficiency in man. *American Journal of Clinical Nutrition*, **24**, 432–43.

Hoffbrand, P. V. (1978). Effects of folate deficiency in man. In *Effect of nutrient deficiencies in man* (ed. M. Reichcigl), pp. 55–68. CRC Press, Florida.

Holt, J., Seaman, J., and Rivers, J. (1975). Famine revisited. *Nature*, **225**, 180–1.

Hoppner, K., Phillips, W. E. J., Murray, T. K., and Campbell, J. S. (1968). Survey of liver vitamin A stores of Canadians. *Canadian Medical Association Journal*, **99**, 983–6.

Hume, E. M., and Krebs, H. A. (eds) (1949). Vitamin A requirements of human adults, Medical Research Council Special Report Series, No. 264. HMSO, London.

Ivanovsky, A. (1923). Physical modification of the population of Russia under famine. *American Journal of Physical Anthropology*, **6**, 331–53.

Jeffrey, D. M. (1982). Pantothenic acid. In *Vitamins in medicine*, Vol. 2 (4th edn) (eds B. M. Barker and D. A. Bender), pp. 69–91. Heinemann, London.

Keneally, T. (1983). *Schindler's Ark* (5th edn). Coronet Books, London.

Keys, A., Brozek, J., Henschel, A., Mickelson, O., and Taylor, H. L. (1950). *The biology of human starvation* (2 vols), University of Minnesota Press.

Kielmann, A. A. and McCord, C. (1978). Weight for height as an index of the risk of death in children. *Lancet*, **i**, 1247–50.

Kleiber, M. (1975). The fire of life. Robert E. Krieger, New York.

Kustner, H. G. V., Whitehorn, R., Wittmer, H., Hignett, V. M., Rawlinson,

J. L., Raubenheimer, W. J. J., and van der Merwe, C. A. (1984). Weight-for-height nutrition surveys in rural Kwazulu and Natal, July 1983. *South African Medical Journal*, **65**, 470–4.

Langan, S. M., and Farell, P. M. (1985). Vitamin E, vitamin A, and essential fatty acid status of patients hospitalised for anorexia nervosa. *American Journal of Clinical Nutrition*, **41**, 1054–60.

Levin, S. L. (1944). Particular form of polyneuropathy with yellow pigmentation of the skin. *Sci. Mem. First Pavlov. Med. Inst.* Leningrad. pp. 63–8.

Lewis, C. F. and Musselman, M. (1946). Observations on pellagra in American prisoners of war in the Philippines. *Journal of Nutrition*, **32**, 549–58.

Leyton, G. B. (1946). Effects of slow starvation. *Lancet*, **251**, 73–9.

Lipscombe, F. M. (1945). Medical aspects of Belsen concentration camp. *Lancet*, **249**, 313–15.

Martorell, R. (1985). Child growth retardation: a discussion of its causes and its relationship to health. In *Nutritional Adaptation in Man*, (eds K. L. Blaxter and J. C. Waterlow), pp. 13–30. John Libbey, London.

Malcolm, L. A. (1970). Growth retardation in a New Guinea boarding school and its response to supplementary feeding. *British Journal of Nutrition*, **24**, 297–305.

McCance, R. A. (1951). The history, significance and aetiology of hunger oedema. In *Studies of undernutrition in Wuppertal, 1946–1949*, pp. 21–82. Medical Research Council Special Report Series, No. 275. HMSO, London.

McCollum, E. V., Simmonds, N., Shipley, P. G., and Park, E. A. (1922). Studies on experimental rickets. XV: the effect of starvation on the healing of rickets. *Johns Hopkins Medical Bulletin*, **33**, 31–3.

MacDonald, I. (1984). Changes in body composition during weight loss. *British Nutrition Foundation Bulletin*, 14–19.

McLaren, D. S. (1974). The great protein fiasco. *Lancet*, **ii**, 93–6.

Magan, A. M., Warsame, M., Ali-Salad, A. K., and Toole, M. J. (1983). An outbreak of scurvy in Somali refugee camps. *Disasters*, 7, 94–97.

Masefield, G. B. (1971). Calculations of the amounts of different foods to be imported into the famine area: emergency subsistence level; temporary maintenance level. In *Famine: a symposium dealing with nutrition and relief operations in times of disaster* (eds G. Blix, Y. Hofvander, and B. Vahlquist), pp. 170–7. Symposium of the Swedish Nutrition Foundation.

Mohammed. (1986). Famine nutrition and health in Sudan. Thesis for M.Sc. (Clin. Trop. Med.), London School of Hygiene & Tropical Medicine, University of London.

Mollison, P. L. (1946). Observations on cases of starvation at Belsen. *British Medical Journal*, **i**, 4–8.

Moore, T. (1957). *Vitamin A*. Elsevier, Amsterdam.

Mount, L. E. (1979). *Adaptation to thermal environment*. Edward Arnold, London.

Murray, M. J., Murray, A. B., Murray, M. B., and Murray, C. J. (1976). Somali food shelters in the Ogaden and their impact on health. *Lancet*, **ii**, 1283–45.

Musselman, M. M. (1945). Nutritional diseases in Cabantuan. *War Medicine*, **8**, 325–32.

Najjar, V. A., Holt, L. E., Johns, G. A., Medairy, G. C., and Fleischmann, G.

(1946). Biosynthesis of nicotinamide in man. *Proc. Soc. Exptl. Biol. Med.*, **61**, 371–4.

Nelson, R. A. (1978). Effects of specific nutrient deficiencies in man: pantothenic acid. In *Effect of nutrient deficiencies in man* (ed. M. Reichcigl), pp. 33–6. CRC Press, Florida.

Netherlands Red Cross Feeding Team. (1948). A report on nutritional survey in the Netherlands East Indies. Van Loon, Hague.

Oomen, H. A. P. C. (1976). Vitamin A in times of disaster. *Ann. soc. belge. Med. trop.*, **56**, 319–23.

Pacey, A. and Payne, P. R. (1985). *Agricultural development and nutrition.* Hutchinson, London.

Pariser, E. R., Wallerstein, M. B., Corkey, C. J., and Brown, N. L. (1978). *Fish protein concentrate: panacea for malnutrition?* MIT Press.

Passmore, R. and Eastwood, M. A. (1986). *Human nutrition and dietetics.* Churchill Livingstone, Edinburgh.

Payne, P. R. (1972). Protein quality of diets, chemical scores and amino acid imbalances. In *International encyclopaedia of food and nutrition* (ed. E. J. Bigwood), Vol. 11, pp. 259–306. Pergamon Press, Oxford.

Petty, C. and Rivers, J. P. W. (1986). Problems in assessing nutritional status in historical studies. *Bulletin of the Society of the Social History of Medicine*, **38**, 77–83.

Pompen, A. W. M., La Chappelle, E. H., Groen, J., and Mercx, K. P. M. (1946). Hongerosteopathie (-osteomalacia) in Nederland. Wetenschappelijke uitgaverij, Amsterdam.

Raica, N. J., Scott, J., Lowry, L., and Sauberlich, H. E. (1972). Vitamin A concentrations in human tissues collected from five areas in the United States. *American Journal of Clinical Nutrition*, **25**, 291–6.

Rivers, J. P. W. (1982). Women and children last: an essay on sex discrimination in Disasters. *Disasters*, **6**, 259–63.

Rivers, J. P. W. and Brown, G. A. (1979). Physiological aspects of shelter deprivation. *Disasters*, **3**, 20–3.

Rivers, J. P. W. and Crawford, M. A. (1974). Maternal nutrition and the sex ratio at birth. *Nature*, **252**, 297–8.

Rivers, J. P. W. and Frankel, T. L. (1981). Essential fatty acid deficiency. *British Medical Bulletin*, **37**, 59–64.

Rivers, J. P. W., Holt, J. F. J., Seaman, J. A., and Bowden, M. R. (1976). Lessons for epidemiology from the Ethiopian famines. *Ann. soc. belge. Med. trop.*, **56**, 345–57.

Rivers, J., Seaman, J., and Holt, J. (1974). Protein requirement. *Lancet*, **ii**, 947.

Robson, J. S. (1972). Disturbances in water and electrolyte metabolism and in hydrogen ion concentration. In *The principles and practice of medicine* (eds S. Davidson and J. McLeod), pp. 213–57, Churchill Livingstone, London.

Rodahl, K. (1949). The toxic effect of polar bear liver, Skifter No. 92 Norsk. Polarinstitutt. I. Kommisjon Hos Jacob Dywbad, Oslo.

Seaman, J. A. (1984). *Epidemiology of natural disasters.* Karger, Basle.

Seaman, J., Holt, J., and Rivers, J. (1974). Harerghe under drought: survey of the effects of drought on human nutrition. Ethiopian Relief Rehabilitation Commission, Addis Ababa.

Seaman, J., Holt, J., and Rivers, J. (1978). The effect of drought on human nutrition in an Ethiopian province. *International Journal of Epidemiology*, 7, 31–40.

Sebrell, W. H. and Butler, R. E. (1938). Riboflavin deficiency in man. *Public Health Reports*, 53, 2282–4.

Shapter, J. (1847). On the prevalance of scurvy. *Lancet*, i, 676–7.

Simon, W. V. (1919). Ueber Hungererkrankungen des Skeletlsystems. *Munch. med. Wochenschrift*, 66, 799–804.

Sinclair, H. M. (1982). Thiamin. In *Vitamins in medicine*, Vol. 2 (4th edn) (eds B. M. Barker and D. A. Bender), pp. 114–67. Heinemann, London.

Smith, D. A. (1947). Effects of dietary restrictions in prisoners of war and internees in the Far East 1942–45. *Proceedings of the XI International Congress on Pure and Applied Chemistry*, 21(7).

Smith and Woodruff (1951). Deficiency diseases in Japanese prison camps, Medical Research Council Special Report Series, No. 274. HMSO, London.

Stein, J. and Fenigstein, H. (1946). Anatomie pathologique de la maladie de famine. In *Maladie de famine: recherches cliniques sur la famine executées dans le Ghetto de Varsovie en 1942* (ed. E. Apfelbaum), pp. 21–77. American Joint Distribution Committee, Warsaw.

Stein, Z., Susser, M., Suenger, G., and Marolla, F. (1975). Famine and human development: the Dutch hunger winter of 1944–45. Oxford University Press.

Szjenman, M. (1946). Le sang dans la maladie de famine. In *Maladie de famine: recherches cliniques sur la famine executées dans le Ghetto de Varsovie en 1942* (ed. E. Apfelbaum), pp. 227–58. American Joint Distribution Committee, Warsaw.

Thurnham, D. I. (1978). Effects of specific nutrient deficiencies in man: riboflavin. In *Effect of nutrient deficiencies in man* (ed. M. Reichcigl), pp. 3–14. CRC Press, Florida.

Valverde, V. (1975). Analysis of some economical aspects relating to prevalence of malnutrition in medium and low income groups of urban and rural areas of Costa Rica. M.Sc. (Human Nutrition) Thesis. London School of Hygiene and Tropical Medicine, University of London.

Vedder, E. B. (1913). *Beriberi*. Bale & Davidson, London.

Vincent, C. P. (1985). *The politics of hunger: the allied blockade of Germany, 1915–1919*. Ohio University Press, London.

Waterlow, J. C. (1972). Classification and definition of protein–calorie malnutrition. *British Medical Journal*, iii, 566–9.

Waterlow, J. C. (1985). What do we mean by adaptation? In *Nutritional adaptation in man* (eds K. L. Blaxter and J. C. Waterlow). John Libbey, London.

Waterlow, J. C. and Payne, P. R. (1976). The protein gap. *Nature*, 258, 113–17.

WHO (1983). Measuring change in nutritional status. WHO, Geneva.

Widdowson, E. M. (1985). Responses to deficits of dietary energy. In *Nutritional adaptation in man*. (eds K. L. Blaxter and J. C. Waterlow). John Libbey, London.

Winick, M. (1979). Editorial annotations in *Hunger disease: studies by the Jewish physicians in the Warsaw Ghetto* (ed. M. Winick). J. Wiley, New York.

Woodham-Smith, C. (1968). *The great hunger: Ireland 1845–9* (2nd edn). New English Library, London.

Yagi, K. (1978). Effect of nutrient deficiencies in man: riboflavin. In *Effect of nutrient deficiencies in man* (ed. M. Reichcigl), pp. 15–32. CRC Press, Florida.

Yavorsky, M., Almaden, P., and King, C. G. (1934). The vitamin C content of human tissues. *Journal of Biological Chemistry*, **106,** 525–9.

3

THE ECONOMICS OF FAMINE

Meghnad Desai

INTRODUCTION

Famines were thought to have disappeared from the scene so much that there was no discussion of them in economics in the twentieth century until recently. The Malthusian threat lost its credibility in the developed Western world and famines were thought unlikely. They were not unknown; in the aftermath of World War I, the revolution and civil war in Russia saw a severe famine leading to a detailed but little-cited study of the influence of hunger on human affairs by the sociologist Pitrim Sorokin (Sorokin 1975). Famines, in view of the West, happened elsewhere. In the period during and after World War II, mass hunger was seen in many occupied countries especially the Netherlands. Again this could be dismissed as an extraordinary event. It was the emergence of the newly independent countries in Asia and Africa that brought issues of development and poverty on the economics agenda.

Even in this area the first two decades of discussion and theorizing were marked by a Panglossian optimism; calculations were made for target growth of per capita income, the required investment and a search was made for resources domestic and foreign. Given enough aid, it was thought the problems of underdevelopment could be overcome. At the onset of the 1970s this optimism broke down. Harvests had already failed in India in two successive years 1965–6 and 1966–7, though no famine deaths were admitted to. (A famine also occurred in China in 1960–1 but this did not become widely known till much later.) The newly-independent Bangladesh was seen quite widely as needing special help – 'basket case' in US State Department jargon but also inviting a concert by former Beatle George Harrison prefiguring the Live Aid of recent years. The year 1972 also saw low harvest in the US and the spectre of raw material shortage caused by too much growth was mooted.

The bad harvests of 1972 were followed by the oil shock of 1973. Suddenly resource shortages were not only a problem for the second and the third world but came home as it were to the first world. The Keynesian optimism was lost and a general search ensued for ways of tackling the age-old problems of scarcity. Unsurprisingly many of the nineteenth-century arguments were revived and refurbished to tackle the recurring problem. Malthusianism became fashionable especially as directed towards the less-developed countries. Hunger was said to be caused by a lack of food and too many mouths to feed. The global perspective that had been quite rightly advanced, by Barbara Ward, for example, was now distorted into the paradigm of a life raft from which some had to be ejected to save the rest. Little imagination is needed to gauge how the third world ranked in those stakes. In macro-economic affairs the revival of *laissez-faire* policy was in tune with these trends.

Against this background, it was Amartya Sen's study of the Great Bengal Famine that offered an alternative but general theory of famines (Sen 1977). This was followed by Sen with further 'applications' of the theory to the famines of the 1970s – Bangladesh, the Sahel, and Ethiopia. This work has been followed up by a number of studies of various regions and historical episodes (for a bibliography, Sen and Dreze 1987). The theory has also been criticized as being either nothing that was not well known or as no more than a minor curiosum in neo-classical economics (Mitra 1982; Srinivasan 1983). Others have sought to extend and generalize Sen's argument (Desai 1984, 1986a; Rangasami 1985).

In this chapter I shall argue that the economic theory offered by Sen has a great potential for extension and development. While the theory could be generalized beyond the case of famines, I concentrate here on the ways in which it needs to be supplemented for a study of famines. A major problem is that the theory lacks a dynamic framework. It needs to be articulated in a system of blocks which bring out the dynamic and simultaneous interdependence of the physical and the economic aspects of famines. Sen's theory will be outlined first and then I look at the criticisms. The following section extends the theory in various ways to meet the critics' objections. In the last section a systems representation will be given in which the dynamic extensions will be made. All along I shall have occasions to mention other studies and examples of the basic approach.

THE THEORY OF EXCHANGE ENTITLEMENTS

The theory of exchange entitlements was advanced by Amartya Sen specifically in an attempt to explain the paradox of one of the most tragic famines of this century – the Great Bengal Famine of 1943. After a careful calculation of the available food supplies, he concluded that the total amount of food available had not declined since the previous years, but despite this there were widespread deaths. Looking further into this, he found that the pattern of mortality was systematically different in rural areas as between people connected in some way with the rice-farming activity and those groups who were in the non-rice and non-agricultural activities. With a further look at how people moved from their usual occupations to new ones as a way of averting the worst consequences of famine, it became clear that the drift was towards agricultural labour or any other activity that would bring them closer to the rice-growing activity. In one sense what happened was that the price of rice had risen relative to that of other goods and services. While this is obvious, it was also true that those not growing rice had less to sell and also that they could purchase less rice with each unit of goods they could sell, i.e. they suffered a loss of quantity as well as adverse terms of trade. But even within the rice-growing sector, those who could appropriate rice directly – as sharecroppers or as labourers paid in terms of rice – suffered less than those who had to 'translate' what they had – money or goods and services – into rice by exchange.

Sen built up his theory further by examining some more recent famines – Bangladesh in 1974, the Sahel famines of the early 1970s, and the Ethiopian famines of 1973 and 1974. In economics the usual practice is to build up a theory *a priori* on axiomatic or other sparse foundations and then subject it to test. Sen's theory was built up with an empirical (not empiricist) approach (see Thompson (1978) for the distinction). The extension of his investigation to African famines enabled Sen to examine a wider class of food and non-food activities. Thus if the main non-food activity in Bengal was fishing, in Africa the pastoral activity was the alternative. In stating his thesis in the book four years after his article, the argument is couched in terms of an attack on the 'Food Availability Decline' (FAD) hypothesis. FAD argues that it is decline in the overall per capita availability of food as a result of the twin pincers of population growth and agricultural stagnation that famines occur. Shifting attention away from this aggregative approach, Sen argues that FAD is neither necessary nor sufficient for famines to occur. In one sense Sen's theory can be seen as a series of arguments demolishing one after another of the monocausal arguments about famines. Thus, for example, droughts and such other extreme natural events are also neither necessary nor sufficient. Famines have occurred with and without FAD, with and without the food grain price rising, and with and without natural disasters occurring.

To understand famines, it is important to place the various economic groups in relation to the food-growing economy. Given their abilities and their endowments as well as the existing set of property rights, etc. one can define the exchange entitlements of each person in terms of the food he can obtain in the 'normal' course of events. If this entitlement is less than the amount required to subsist then the person will starve. One could think of poverty as endemic low entitlements. This happens because the poor have low-ownership entitlements; whatever the configuration of relative prices, they can obtain inadequate amounts of food (or in a more general context a basket of goods) with what they have to offer. But for those who may have normally adequate entitlements in terms of what they own, famines are situations in which their exchange entitlements to food can suddenly shrink below starvation levels. It is necessary, therefore, to examine the nature of entitlements both as they relate to individuals and as compared between different groups of individuals. (In what follows I rely on Desai (1984).)

Each person, no matter how poor, has some endowments. They may consist of personal attributes – age, sex, race, height, weight, and more elusive personal qualities such as charm, beauty, etc. In terms of economic measures every person has at least his capacity for work – his labour power – unless too young or too old or infirm or severely handicapped. Others may have additional endowments – land, money, durable goods, financial assets, etc. There are certain legal rules which define what can and what cannot be owned. Thus a slave does not have any endowments which he can call his own except the personal attributes which may have made him more marketable. Serfs were not free to dispose of their labour power as they chose. Modern societies do not permit slaves as part of one's endowment.

Given the endowments, there are various ways of converting them into goods and services which constitute consumption baskets. Some endowments can be directly traded into goods, e.g. money is such a universal solvent. Others have to be put into a production process – seeds or ploughs – so they can yield output – things – which can be converted into other goods. Labour power has a dual status in this scheme. It can be directly sold for wages which can be converted into consumption goods. It can also be engaged into a production process along with other inputs which will yield output which again can be sold. All these exchanges whether of labour directly or of labour and inputs via output presume a market and they are all mediated via money in almost all modern economies no matter how 'less developed'. There also has to be sufficient demand for these 'vendible commodities' (to use Adam Smith's expression) for one to be able to obtain other goods for it.

Now, given one's endowments there are various ways of converting them into goods. If I have land, I could sell it outright and live off the sum of money with or without additional interest earned by investing the capital sum. I could alternatively rent the land out and live off the rent. Or I could give it to a sharecropper for half the output though I may also have to provide some of the inputs. I may, lastly, cultivate the land myself with or without hired labour.

Thus, from the ownership of one asset – land, here are various possibilities each of which will convert into a basket of consumption goods. Selling the land or renting it at a money rent will yield me sums of money which I could trade for consumption goods at the prevailing retail prices. If, however, I have let it for a rent in terms of a portion of the output, the situation is similar to sharecropping and to own cultivation. I have to *sell* the output and then buy consumption goods with the money revenue earned from the sale. Here again it is crucial for the probability of my experiencing starvation whether I grow edible or non-edible crops. For the former at least some portion of the output can be directly consumed and the price does not enter into the question. For non-edible crops there has to be a sale before there can be a purchase.

Now all the various combinations of consumption bundles I can obtain for my land traversing these routes is what Amartya Sen calls my 'exchange entitlements'. If I owned only my labour power, then I have the choice of working for a wage as an agricultural (landless) labourer, or as a non-agricultural labourer, or perhaps a sharecropper. The labourer and the sharecropper are similar because they own nothing but their labour power. They may turn out to have different exchange entitlements, however.

In 'normal' times, the wage of the worker will be enough to provide subsistence, if not more. Thus the subsistence bundle is within his exchange entitlement. The same, let us assume, is true for the sharecropper, the tenant, and the landlord. Now, if for some reason – drought, war, black-market hoarding – the price of grain were to go up, some of the participants will find their entitlement shrinking. This may be so drastic that they may not even get a subsistence bundle. They may starve.

But even within this context of shortage, caused either by demand shock (war, hoarding) or supply shocks (drought, floods, locusts), we must examine the mode of income receipt to appreciate the uneven incidence of famine on different people. Take as an extreme case two landless labourers – one paid money wages, and one paid in kind. Let us assume that the wage is fixed before the harvest *ex ante* at Rs1000 for the season, or 100 kg of grain. If the rains fail, or due to some other reason harvests are very low, the output will be below expectations. If this is a widespread phenomenon, grain prices may rise. If 50 kg is a subsistence bundle, a price rise of grain of more than 100 per cent would push the worker paid in money wages below the subsistence line. Thus his exchange entitlements shrink due to *relative price changes*. The worker paid in grain will then be that much safer (assuming the employer does not arbitrarily renege on the contract to pay him 100 kg).

Take now a sharecropper and a worker paid in money wages. Let us say the sharecropper may normally expect 150 kg of grain – one half of normal output of 300 kg. Again if output were to be anywhere lower than 100 kg due to, say, drought, then the sharecropper would be pushed below subsistence. So he will suffer not so much from relative price change as from a pure quantity change.

Take lastly a cattleman. The nomadic herdsmen of the Sahel raise livestock and take it from one place to another for grazing. They sell cattle or dairy products to buy grain. A drought would mean decline in cattle output (in terms of weight) and/or in dairy output. If, in addition, grain prices have also risen, then a cattleman would suffer relative price loss (grain is more expensive in terms of dairy products or meat) and from output loss (leaner cattle/less milk). His entitlement of grain would thus shrink due to both these reasons and he will be doubly vulnerable, as it were, to starvation.

Thus, whatever the macro-economic dimensions of food shortage, the micro-economic incidence of starvation would depend on how individual households are placed in terms of their endowments, and through these endowments, in their exchange entitlements. To some extent these endowments relate to ownership of the means of production as in the case of sharecropper/labourer but they also depend on *access* to the means of production. But here again – ownership or access to the means of production are not sufficient to define the likelihood of entitlement loss though they may serve to determine the class to which an individual belongs. If you have two modes of production coexisting, say a nomadic one with livestock as the principal means of production and an agricultural one, the class position of a rich cattle owner does not render him immune from entitlement loss if the cattle/grain price ratio changes. Thus entitlement leads to a more complex classification than a Marxian-class analysis.

CRITIQUES

The entitlement approach to famines has not gone without criticism. A major criticism has been that in some sense what the approach says is obvious and well known. Thus in an early review of the book, Ashok Mitra was dismissive of Sen's contribution saying that after all this is what our grandmothers always knew about famines (Mitra 1982). Srinivasan in another review has taken the view that the only thing Sen says is that the real wage, in terms of grain, falls when the price of grain goes up. The purchasing power having shrunk, people buy less than required for survival. Hence they starve. 'The "entitlement approach" is a fancy name for elementary ideas fairly well understood by economists, though not necessarily by policy makers' (Srinivasan 1983, p. 200). Srinivasan also points out that the shifts in purchasing power have to be drastic and sudden for the analysis to have any bite; 'A less drastic and gradual shift (in real purchasing power) would have given enough time for authorities to take appropriate action and individuals to adjust on their own' (Srinivasan 1983, p. 201). Once we take such dynamic considerations into account 'the role of price expectations and speculative changes in food stock have to be analysed for the dynamics of price–wage movements to be understood' (Srinivasan 1983, p. 201).

A criticism of a different sort has been made by Amrita Rangasami (1985). Her point is that famine and starvation are not results of sudden and drastic collapses in purchasing power or food supply. There is a process which starts much before which has already rendered some groups more vulnerable than others. Rangasami's points about the dynamic process that ends in starvation can be summarized as follows (all quotes from Rangasami 1985, p. 1748):

(a) ... that starvation is a process and that it is long drawn, hardly sudden;
(b) ... that the biological process has a socio-economic dimension: that such a process has clearly marked phases. The phases correspond with biological changes and deterioration in the health of the affected community and socio-economically by transfer of assets from victim to beneficiary. The socio-economic process is completed with the loss of all the victim's assets including his ability to labour;
(c) ... that the state does not appear to intervene until the third and irreversible phase;
(d) ... that the perceptions of famine we have today only relate to the terminal phase and not the entire process. Consequently, they have a limited validity. I will, therefore, conclude that the definitions in use insofar as they hinge on the elevation of mortality may be set aside. Consequently, Sen's work which is based on such a definition is inadequate.

Rangasami's focus is wider than the narrowly economic theoretic criticisms of Srinivasan. She is concerned both with the biological process whereby there is a dynamic interplay between inadequate food intake and undernourishment which may eventually lead to starvation. Citing various nutritional and clinical studies on starvation, she says, 'Together these studies demonstrate that the individual passes from a well nourished stage through successive stages of starvation. Far more significant that the body can adapt itself at a low equilibrium – a plateauing effect that can endure for weeks to several months ... ' (Rangasami 1985).

Rangasami's consequent concern is that famines are defined too partially. Even in normal times some poor individuals may die of starvation: 'Famine is a condition that affects large numbers of people within a recognisable, spatial unit such as a village, a country or an entire geographical region.' Citing studies of the famine in the Netherlands during Nazi occupation, she lists the three stages of famine as dearth, famishment, and morbidity. Within this dynamic process there is a steady loss of the assets of the victim and a deterioration in the terms at which he trades with the beneficiary. The movement from dearth to morbidity is via famishment. Famishment indicates movement from a state of dearth where there is still some hope of returning to the original stage before morbidity when such hope is abandoned. It is at this stage that 'strategies to prevent death ... become imperative. These can include acceptance of slavery, conversion to other religions, permanent migration as indentured labour. What is significant is that a large number of families resort to one identical stratagem' (Rangasami 1985, p. 1750).

Rangasami also points out that an examination of Indian Famine Codes and actual practice relief confirms that Sen's entitlement approach was understood by the administrators. That does not mean that they knew of its formal structure but the logic of looking at the variety of factors determining the exchange entitlements of individuals. Thus they looked not only at food shortages, but also unemployment and real wages. They devised work programmes to generate purchasing power for those most likely to suffer from high food prices, 'The relationship between drought and famine was not so much a reduction in total food output as a decline in the level of employment' (Rangasami 1985, p. 1798).

I have described Rangasami's critique at length because it introduces crucial non-economic factors while broadening the economic factors to include dynamic considerations. In spirit, it is not hostile to the entitlement approach but wishes it to move beyond the static, one-period consideration. In this sense one can say that while Sen's theoretical framework is static as well as partial equilibrium, in his concrete studies of the four famines many of the other considerations do play a role. It is important, however, for our purposes, to supplement and extend Sen's framework to take account of some of the criticisms.

EXTENSIONS OF THE ENTITLEMENT APPROACH

THE ENTITLEMENT APPROACH: SOME SIMPLE ANALYTICS

There are two strategies which I shall pursue in extending the entitlement approach. In a narrow economic theoretic framework I shall explore ways of clarifying and extending the model. Taking Sen's theoretical framework in his book and his QJE article (Sen 1981*b*), I shall extend it to a general equilibrium argument demonstrating the asymmetry between food markets and other markets. Somewhere in the course of a famine, food becomes a pivotal commodity, displacing all other commodities to a peripheral status. How is one to account for this in a formal model? Having done that in the context of a two person–two good economy, I shall next ask how we can incorporate the dynamic considerations emphasized by Rangasami.

To begin with let us look at the diagram used by Sen in his book and in his article in the *Quarterly Journal Economics*. This is given here as Fig. 3.1. An individual requires OA amount of food to avoid starvation. The area DAE is thus outside the starvation set. He has an initial endowment of $x^\star$ representing a combination of food and non-food commodities. Initially the rate at which he can exchange his endowments for food is given by the price slope p and the line AB (the slope is tan of angle BAO). It is clear that if he is at $x^\star$, he can, by converting all his non-food endowments into food, get beyond OA and be outside the starvation set. Now if the endowment set were to drop (quantity shock) from $x^\star$ to $x^\star$, the individual would get food less than OA. But even without this quantity shock, if the price of food were to rise from p to $p^\star$ (tan of angle CAO), then even with $x^\star$, the individual will get less than OA. Famines can thus happen either with or without a food rise or with or without a quantity shock. Of course in actual famines, individuals often suffer from both quantity and price shocks.

It will be helpful to begin by considering a modification of Fig. 3.1. A

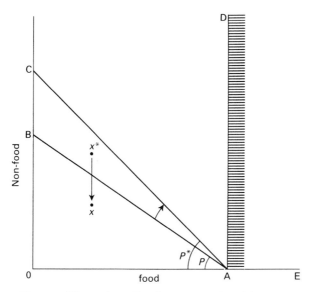

Fig. 3.1. Illustration of endowment and entitlement.

person may either offer labour services as non-food endowment or another commodity. In either case the initial endowment will be a point on the non-food axis rather than in the interior, i.e. a combination of food and non-food. For someone who supplies labour services, a point such as L_1 represents the number of hours he can work, the price slope then representing the real wage in terms of food. Let this food commodity be rice. Fig. 3.2a is an adaptation of Fig. 3.1. For a labourer supplying labour power above OB at price (i.e. real wage) p, there is no fear of starvation. If

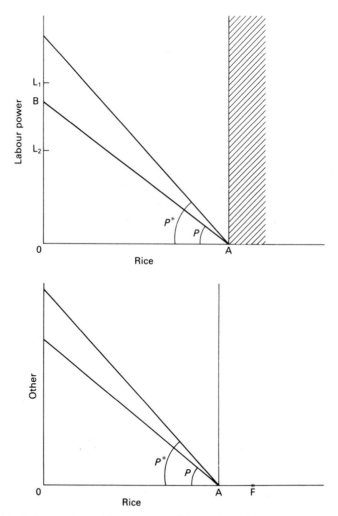

Fig. 3.2(a). Labourer's entitlements. **(b).** Farmer's entitlements. Note: farmer's endowment is independent of price variation.

prices move to $p\star$ (real wage falls) or if he is debilitated and cannot work more than OB hours, then he will begin to face the process which will eventually culminate in starvation.

By contrast, a rice farmer will have his endowment as the output of rice. As long as this is beyond OA he will not starve (Fig. 3.2b). Typically, he may wish to be much beyond this if he is hiring a labourer. But notice that if the rice farmer's output is at some point beond OA, he may not wish to purchase any non-food commodity, i.e. not hire labour. A shortage of food endowment will not only impoverish the farmer but also shrink the demand for other products and services by the farmer.

A GENERAL EQUILIBRIUM EXTENSION

It is obvious, therefore, that we should examine food demand and food supply together with demand and supply of other commodities that non-food producers have to exchange for food. Let us suppose that there are two such commodities, rice and fish, and two individuals, the fisherman (A) and the farmer (B). Let us look at the possibility of a famine-type entitlement failure in this static but general equilibrium framework. (I am drawing here on Desai 1986a.)

I assume, not unreasonably, that both the individuals regard rice as their basic food and fish as something extra worth having but not crucial to survival. Indicate the amount of rice required as a minimum by each individual as $X_{1A}^{\star}$ and $X_{1B}^{\star}$. No such minimum amount is necessary for fish but fish yields utility to the consumer. The formal representation of such a utility function is:

$$U_i = \alpha_1 \log(X_{1i} - X_{1i}^{\star}) + (1 - \alpha_1) \log X_{2i}, \quad i = A, B \qquad (1)$$

Each individual derives utility from rice consumption if he gets more than $X_{1i}^{\star}$ not otherwise. The weights α_1 and $(1 - \alpha_1)$ attach to rice and fish consumption. Now the rice farmer's income will be $p_1 \cdot X_1$, X_1 being total output and p_1 the unit price, and the fisherman's income will be $p_2 \cdot X_2$, X_2 being fish output.

Given the form of the utility function in (1), it is easy to derive the demand functions of the farmer and the fisherman for rice and fish. (Formally this is done by maximizing U_i subject to the constraint that total expenditure of each individual does not exceed his income.) We have:

$$X_{1A} = X_{1A}^{\star} + \frac{\alpha_1 (p_2 \times X_2 - p_1 X_{1A}^{\star})}{p_1} \qquad (2a)$$

$$X_{2A} = (1 - \alpha_1) \frac{(p_2 X_2 - p_1 X_{1A}^{\star})}{p_2} \qquad (2b)$$

$$X_{1B} = X_{1B}^{\star} + \alpha_1(X_1 - X_{1B}^{\star}) \tag{2c}$$

$$X_{2B} = (1 - \alpha_1)\frac{p_1(X_1 - X_{1B}^{\star})}{p_2} \tag{2d}$$

Each equation gives us the demand of an individual for a commodity given his income and prices except in the case of the farmer's demand for rice (X_{1B}). Here note that prices play no role whatsoever. The farmer decides his consumption of the pivotal commodity at the outset. He consumes above his subsistence requirement a proportion of his total output above his subsistence requirement, i.e. trivially rewrite (2c) as:

$$(X_{1B} - X_{1B}^{\star}) = \alpha_1(X_1 - X_{1B}^{\star}) \tag{2c'}$$

This, however, means that the fisherman only gets the excess above X_{1B} i.e. ($X_1 - X_{1B}$) as a possible quantity he can purchase. There is no guarantee at all that this quantity will suffice for him i.e. that $X_{1A}^{\star} < (X_1 - X_{1B})$. Being away from the production of the foodstuff, he cannot guarantee that he will be able to obtain his subsistence requirement in exchange. Another way to put this is that in equation (2a) there is nothing to guarantee that $X_{1A} > X_{1A}^{\star}$ i.e. that $(p_2 X_2 - p_1 X_{1A}^{\star}) > 0$. It will depend on the amount of marketed surplus ($X_1 - X_{1B}^{\star}$) that the farmer will release and the resulting terms of trade.

As far as the terms of trade are concerned, we can explicitly solve them out given the output quantities X_1 and X_2. We have

$$\frac{p_1}{p_2} = \frac{\alpha_1 X_2}{(1 - \alpha_1)(X_1 - (X_{1A}^{\star} + X_{1B}^{\star}))} \tag{3}$$

If for some reason enough is not produced to cover the subsistence requirement for *both* the individuals $X_1 < (X_{1A}^{\star} + X_{1B}^{\star})$, the price ratio turns negative, i.e. trade breaks down. Of course, in case output of rice is actually below the subsistence requirements of the two individuals, what will happen is that the farmer will grab X_{1B} and leave the fisherman $X_1 - X_{1B}$.

This is a simple and stylized example to show how if we require minimum food for subsistence and let one person (or group) specialize in food production, then that person (or group) occupies an asymmetric position in exchange. Their food entitlements are always guaranteed and the other groups, not in food production, become marginalized. The price ratio p_1/p_2 will be steeper as food output above subsistence needs falls. Let us look at some examples.

We consider three examples. The first example sets up a 'normal' situation to set the scene and the following ones are variants with differential drops in output of the two commodities.

Example 1: assume that $\alpha_1 = 0{\cdot}6$, $X_{1A}^\star = X_{2B}^\star = 500\,\text{kg}$. Outputs are $X_1 = 3000\,\text{kg}$, $X_2 = 1000\,\text{kg}$. Put $p_2 = 1$ by assumption. Then we get:

$$p_1/p_2 = 0{\cdot}75, \qquad X_{1A} = 1000\,\text{kg}, \qquad X_{1B} = 2000\,\text{kg}$$

Example 2: let the output of rice be halved, $X_1 = 1500\,\text{kg}$. Everything else as above:

$$p_1/p_2 = 3{\cdot}0, \qquad X_{1A} = 400\,\text{kg}, \qquad X_{1B} = 1100\,\text{kg}$$

Thus the price ratio quadruples because though output declines by 50 per cent, marketed output in the new situation is a quarter of what it was before. The farmer gets 55 per cent of his former consumption but the fisherman only 40 per cent.

Example 3: let both outputs be halved, $X_1 = 1500\,\text{kg}$, $X_2 = 500\,\text{kg}$:

$$p_1/p_2 = 1{\cdot}5, \qquad X_{1A} = 400\,\text{kg}, \qquad X_{1B} = 1100\,\text{kg}$$

This is a strong 'invariance' result whereby the rice farmer's rice consumption is unaffected by the fall in the output of fish and the consequent fall in the relative price of rice. The relation between the two individuals is not symmetric but asymmetric and 'recursive'. The rice economy determines the rice allocation independently of the fish economy.

BOOM FAMINES AND SLUMP FAMINES

But, of course, famines can arise without the output of the pivotal commodity going down. Amartya Sen has distinguished between boom famines and slump famines. The Great Bengal Famine of 1943 was caused by the demand shift due to the wartime needs of the military and the consequent upturn in the activity it caused in urban areas. Thus without the fish/rice output balance changing in any way, prices could still go up due to a demand shock.

To demonstrate this, we should ideally have a more elaborate model with several income groups all interacting. One such model is presented in Appendix B of Poverty and Famine (Sen 1981a). This has five different classes and is designed to show the uneven incidence of a rise in the minimum ration on the different economic groups. We can do the same in a slightly amended version of the model above. Thus assume that apart from the two individuals A and B there is an outside 'urban' economy. From this economy, two types of shocks can be felt. One is that a sudden influx of purchasing power denoted m can be injected to buy rice. This is a demand shock. (Technically, one could think of m as an unexpected deviation from normal demand which is already accommodated in the economy.) On the supply side, a shock can come from forced requisition of rice. Such things happen during civil wars. Call this shock $\hat{X}_1$.

Without much ado, we can write down the modified equation (3) as a result of these two shocks

$$\frac{p_1}{p_2} = \frac{\alpha_1 X_2 + m}{(1 - \alpha_1)(X_1 - \hat{X}_1 - \Sigma X_1^\star)} \tag{3a}$$

Thus the effect of m and of $\hat{X}_1$ is to raise the price ratio p_1/p_2. But notice again that the incidence of the two shocks is very different on the farmer as against the fisherman. Let us take the two shocks in turn

(a) Demand Shock: $m > 0$, $\hat{X}_1 = 0$

It is quite clear from our earlier analysis that the farmer's rice consumption stays as it was in equation ($2c'$). The burden of the extra demand falls on the fisherman. Take again Example 1 but say that $m = 1000$ (money units). Then the resulting equilibrium is:

Example 1m:

$X_1 = 3000,$ $X_2 = 1000,$ $p_1 = 0.6,$ $p_2 = 1,$ $m = 1000$

$p_1/p_2 = 2.0,$ $X_{1A} = 500,$ $X_{1B} = 2000,$ $X_{1m}(= m/p_1) = 500$

Example 2m: as Example 2 above but with $m = 1000$

$p_1/p_2 = 8.0,$ $X_{1A} = 275,$ $X_{1B} = 1100,$ $X_{1m} = 125$

Example 3m: as in Example 3 above but with $m = 1000$

$p_1/p_2 = 6.5,$ $X_{1A} = 247,$ $X_{1B} = 1100,$ $X_{1m} = 153$

The demand shock examples illustrate several things. First, if examples 1 and 1m were two successive years where everything else was unchanged except an injection of extra demand we see how the fisherman's consumption of rice is cut to the bare bones and he suffers the entire incidence of the extra demand. This occurs without any decline in rice output. The next interesting contrast is between examples 2m and 3m as compared to examples 2 and 3. In the case of 2 and 3, there was invariance in the rice allocation as between the farmer and the fisherman, although fish output halved. Now when fish output halves, in the presence of outside demand of the nominal magnitude, the fisherman is worse off. He has now suffered quantity shock and price shock.

While it is not strictly legitimate to think of these examples as occurring successively through time, one can imagine that a transition from the world of example 1 to that of example 3m, sees the fisherman going from 1000 to 247 in his rice entitlement due to (i) drop in rice output, (ii) drop in fish output, (iii) presence of outside demand.

(b) Supply Shock $\hat{X}_1 > 0$, $m = 0$

The supply shock story is straightforward in as much as it acts in a similar way to a drop in output. There may be some difference if the grain is collected at the farm gate and the farmer is left with only $X_1 - \hat{X}_1$, from which to meet his own requirement X_{1B}^* as well as sell some output. On the other hand if $\hat{X}_1$ is collected from the marketed output, i.e. from $(X_1 - X_{1B})$ the effects would be wholly on the fisherman.

Example 1x:

Take $\hat{X}_1$ as a lump sum tax of 500. $X_1 = 3000$, $X_2 = 1000$, $\alpha_1 = 0.6$ as in Example 1 before. Then by $(2c')$ $X_{1B} = 1700$, $X_{1A} = 800$ and $\hat{X}_1 = 500$. If on the other hand it is collected from marketed output we could have $X_{1B} = 2000$, $X_{1A} = 500$, $\hat{X}_1 = 500$. The price ratio p_1/p_2 is the same in the two cases which is a consequence of the equation $(3a)$ above, i.e. $p_1/p_2 = 1$. But given the independence of the farmer's consumption of relative price, the two different methods of procurement have different effects.

Example 2x:

Again with $\hat{X}_1 = 500$ and the values of Example 2, we get

    Direct requisition $X_{1B} = 500$, $X_{1A} = 500$, $\hat{X}_1 = 500$

    Indirect requisition $X_{1B} = 1100$ with not enough left over to even collect
        $X_1 = 500$ let alone have $X_{1A} > 0$.

Examples of such direct requisition are not totally unrealistic. The Soviet famine of 1920 had as one of its causal factors the policy of direct requisition of food grains. At a time when food output was only 60 per cent of its 1913 value (industrial output barely 20 per cent), the policy of direct requisitioning from farmers to feed the urban workers led to a tremendous hardship, and reportedly the death of between 2 to 3 million in 1920 (Jasny 1972, p.12; Sorokin 1975). It was on recognition of the cost of such direct requisition policy that Lenin decided to shift to market-based agricultural requisition policy (Lenin 1971; Desai 1971).

    There are various directions in which we can extend such analysis further. Thus, one well-established sign of distress is when people start selling their assets in order to obtain cash to buy food. Such distress sales only result in the assets not bringing in as much as they would in normal times because those who have the cash to buy such assets have a strong bargaining position. But the presence of financial assets does modify the stark results given in examples 1, 2, and 3, especially the last two. Having established the foundations in our earlier equations, the extension to the case where the two individuals have financial assets is straightforward.

    Let us see this by a simple extension of our example above. Let m_A be the value of assets of the fisherman, m_B of the farmer. Then it is easy to see that the equations $(2a)$–$(2d)$ above are modified as:

$$X_{1A} = X_{1A}^{\star} + \frac{\alpha_1(p_2 X_2 + m_A - p_1 X_{1A}^{\star})}{p_1} \qquad (4a)$$

$$X_{2A} = \frac{(1 - \alpha_1)(p_2 X_2 + m_A - p_1 X_{1A}^{\star})}{p_2} \qquad (4b)$$

$$X_{1B} = X_{1B}^{\star} + \alpha_1(X_1 - X_{1B}^{\star}) + \alpha_1 \frac{m_A}{p_1} \qquad (4c)$$

$$X_{2B} = (1 - \alpha_1)\frac{p_1(X_1 - X_{1B}^{\star})}{p_2} + \frac{(1 - \alpha_1)m_A}{p_2} \qquad (4d)$$

We see that each person's demands are augmented by the presence of assets. Also now the farmer's demand for rice (X_{1B}) is no longer insensitive to prices. The price level itself is influenced by the assets and equation (3) has to be modified to:

$$\frac{p_1}{p_2} = \frac{\alpha_1(X_2 + (m_A + m_B))}{(1 - \alpha_1)(X_1 - (X_{1A}^{\star} + X_{1B}^{\star}))} \qquad (3a)$$

We can again work out three examples with financial assets added:

Example 1M: all the values of example 1 but with $m_A = m_B = £1000$. Then:

$$p_1/p_2 = 2.25, \qquad X_{1A} = 734, \qquad X_{1B} = 2266$$

Example 2M: all the values of example 2 with $m_A = m_B = £1000$. Then:

$$p_1/p_2 = 9.0, \qquad X_{1A} = 434, \qquad X_{1B} = 1166$$

Example 3M: all the values of example 3 with $m_A = m_B = £1000$.

$$p_1/p_2 = 7.5, \qquad X_{1A} = 420, \qquad X_{1B} = 1180$$

Several things are noteworthy about these examples. Thus, the invariance of the farmer's rice consumption to prices is now modified but note that by assigning equal money balances to the two individuals, we have tilted the entitlement ratios even more in favour of the farmer than before. When the rice output falls, again his relative advantage improves with assets rather than without. Finally in example 3M, we see that a fall in fish output hurts the fisherman even further in terms of food entitlement than when fish output was normal. Thus example 3M is an illustration of the adverse price and quantity effects persisting even when the individuals have financial assets.

These are again stylized examples, but they serve our purpose. More realism can be added by allowing each individual to demand money balances as well as the two commodities, by linking money balances to past savings and by letting interest rates exert some effect but I will not follow that road here.

One further dimension along which we can explore the implications of the model is to ask how sensitive the outcome is to the parameters α_1 and $X_{1i}^\star$, the minimum quantities. Let us concentrate on the fisherman. From (2a) we see that if the quantity in the parenthesis $(p_2 X_2 - p_1 X_{1A}^\star)$ is negative, the fisherman will get less than his subsistence. For what ranges of the parameters is this likely? Clearly, it is sufficient that:

$$X_2 / X_{1A}^\star < p_1 / p_2$$

for $X_{1A} < X_{1A}^\star$. Exploring this further by substituting equation (13) for p_1/p_2, we get the condition that:

$$\frac{(1 - \alpha_1)}{\alpha_1} < \frac{X_{1A}^\star}{(X_1 - \Sigma X_{1i}^\star)} \tag{5}$$

is sufficient for $X_{1A} < X_{1A}^\star$. Notice that (5) is independent of the value of X_2 as long as it is positive. The fisherman will starve if (5) is fulfilled. In (5), the denominator is the surplus above the farmer's and the fisherman's subsistence and the numerator is of course the fisherman's subsistence. It is clear that the higher α_1 is, the more likely, *ceteris paribus*, that the fisherman will get less than his subsistence.

AN ALTERNATIVE MODEL

The above exercises with a simple general equilibrium model have brought out various aspects of the theory of entitlement. In particular, the asymmetric position of the food producer *vis-à-vis* the person outside the pivotal commodity production is dramatically brought out. The influence of an aggregate demand shock as well as supply shock can be explored in a way that is simple and transparent. The drawback is that we rely on a rather restrictive utility function such as equation (1). While it is patently realistic to require that people need minimum quantities before they derive any utility from consumption, this may yet be found to restrict the generality of our results. We proceed therefore to explore an alternative formulation which preserves the asymmetry of the two commodities but does not require minimum quantities of rice to be consumed.

Instead of equation (1) as our utility function, let us posit the following:

$$U_i = X_{1i}^\alpha (\beta X_{1i} + X_{2i})^{1 - \alpha} \tag{5}$$

Equation (5) is a modified version of the well-known Cobb–Douglas form familiar to economists. The presence of $\beta \neq 0$ is the new element here. As long as $\beta > 0$, the commodity X_1 will be more important in the consumer's mind than X_2. Indeed U_1 has the property that the consumer can go without X_2 and have $U > 0$ but for $X_1 = 0$, $U = 0$. Thus X_2 is inessential and X_1 is essential but in a different sense than in (1) above where both were essential but X_1 was required to be in excess of a minimum quantity.

Like (1), equation (5) is easy to manipulate. Corresponding to (2a)–(2d), we can derive the demand conditions for the two person–two good economy:

$$X_{1A} = \alpha \frac{p_2 X_2}{(p_1 - \beta p_2)} \tag{6a}$$

$$X_{1B} = \alpha \frac{p_1 X_1}{(p_1 - \beta p_2)} \tag{6b}$$

$$X_{2A} = \frac{X_2((1 - \alpha)p_1 - \beta p_2)}{(p_1 - \beta p_2)} \tag{6c}$$

$$X_{2B} = \frac{p_1 X_1((1 - \alpha)p_1 - \beta p_2)}{(p_1 - \beta p_2)} \tag{6d}$$

The structure of demand equations (6a)–(6d) sets some limits to the movements of the relative prices. Thus for the X_1 demand equations it is easy to see that we require the denominator to be positive so that X_{ij} can be positive. Hence $p_1/p_2 > \beta$ is a necessary condition. But given that $p_1/p_2 > \beta$, for demand for X_2 to be nonnegative we require $p_1/p_2 > \beta/(1 - \alpha)$ as can be seen from (7c) and (7d). Since $0 < \alpha < 1$, it is only the latter condition that is of interest.

Thus in the asymmetric world described to us by equation (5), there is a limit above which the price of X_2 cannot rise relative to that of X_1. When the ratio p_1/p_2 falls to $\beta/(1 - \alpha)$, consumers stop buying X_2. At this stage people buy only X_1 but given that no one buys X_2, the producer of X_2 has no income to spend on buying X_1. In this sense, a decline in the output of the peripheral commodity will not benefit its producers. They will lose their entitlements to food (X_1) if the price of their product rises too high. Thus, adverse effects are not compensated by favourable price effects for the producers of the peripheral commodity. The asymmetry is reinforced but in a different way.

We can also work out the equilibrium price ratio corresponding to (3), for the new model. We get:

$$p_1/p_2 = \frac{\beta}{(1 - \alpha)} + \frac{\alpha}{(1 - \alpha)} \frac{X_2}{X_1} \tag{7}$$

Thus the price ratio can fall to the limit when demands for X_2 are zero only if the output of X_2 is zero, since otherwise X_2 and X_1 should be positive.

It is instructive to work out numerical examples on the lines of examples 1 to 3 above. We label them 1, 2 and 3.

Example 1: $\alpha = 0.6$, $\beta = (1 - \alpha)$, other values as in Example 1

$\quad p_1/p_2 = 1.5$, $\qquad X_{1A} = 545$, $\qquad X_{1B} = 2455$

Example 2: α and β as in 1 and other values as in Example 2

$\quad p_1/p_2 = 2.0$, $\qquad X_{1A} = 375$, $\qquad X_{1B} = 1125$

Example 3: α and β as in 1 and other values as in Example 3

$\quad p_1/p_2 = 1.5$, $\qquad X_{1A} = 545$, $\qquad X_{1B} = 2455$

It is clear that in this economy the relative entitlements of rice (X_1) are determined by the ratio of outputs of X_1 to X_2. From $(6a)$ and $(6b)$:

$$\frac{X_{1B}}{X_{1A}} = \frac{\alpha}{(1 - \alpha)} + \frac{\beta}{(1 - \alpha)} \frac{X_1}{X_2} \tag{8}$$

Notice that in $(6b)$ the farmer's demand for rice is no longer independent of the relative prices.

This makes a substantial difference to the farmer's supply response when the relative price changes. Thus if we compare 1 to 2, the proportion of rice marketed (X_{1A}/X_1) goes up from 12·5 to 36·3 per cent as output drops by 50 per cent and the price goes up by a third. By contrast in examples 1 and 2, the proportion marketed falls from 33·3 per cent to 26·6 per cent. The presence of $X_{1i}^{\star} > 0$ and $X_{2i}^{\star} = 0$ in equation (1) thus strongly shapes the supply response. In the second model there is still an asymmetry since there are limits to the price ratio p_1/p_2 in the downward direction but not in the upward direction. Of course, the strength of this asymmetry will depend crucially on the relation of α to β.

The purpose of these two models has been pedagogic. In as much as the entitlement approach has been criticized as either not novel or mistaken in the context of the neoclassical economy theory, I have been at some pains to point out that it can be formulated in fairly general terms and interesting conclusions can be derived from it without violating postulates of neoclassical economics. The case of minimum consumption requirements has been well known to cause problems of existence of equilibrium (Koopmans 1957). While this is well known it has not been sufficiently explored in the literature. What does it precisely mean to say equilibrium does not exist? We see in example 2 that a positive and finite price ratio can be worked out as long as X_1 exceeds the sum of the minimum requirements $X_{1i}^{\star}$. This however leads to a peculiar sort of equilibrium where one of the parties gets less than the minimum requirements. In one sense this violates the utility function but in another sense this represents the best that individual A can do. The fisherman derives 'infinitely negative utility' when $X_{1A} < X_{1A}^{\star}$. But then it is realistic to say that he would rather have a half a loaf than none. In Rangasami's terms he is in the process of famishment. One would need to reformulate the model with not only utility maximization but some food security motives to put some realism into the choice behaviour. Thus it is not utility of consumption but the reproduction of daily existence that may have to become the objective function. But this is clearly an area for future research.

A major omission in the entitlement approach is the dynamics of situations which start from a drought or even normal rainfall but deteriorate into famines. This is not to say that we lack descriptive studies of famines which bring these dynamic aspects out. It is the lack of a theoretical framework that is at issue. Here I have no new analytical models to propose but it is appropriate to list some considerations which may be tackled in the future. One dynamic consideration is the state of the overall economy/ecology within which a trigger event such as drought takes place. If the economy is already weakened by such previous episodes, it will have limited reserves for coping with such a crisis. It is also very likely that if there is no quick and adequate relief, one year's crisis may affect future output. This is especially true in livestock operations where, if some cattle die during a famine, restocking may take several years due to the usual gestation lags. The institutional structures of land ownership and developments in non-agricultural parts of the economy would also be important. Thus while the Great Bengal Famine of 1943 occurred in absence of a decline in food availability, the enormous damage it caused can only be understood if seen in the context of a secular decline in Bengal's economy. The highly concentrated land ownership, the multiple layers of subinfeudation, and the stagnation in the region's industrial economy in the inter-war years had already made Bengal's economy/ecology vulnerable to a shock (Greenough 1982).

Another sort of dynamic consideration is at the individual level. As Rangasami points out, there is a gap between the first shortage – dearth – made good by selling financial assets and the final stage of starvation. Nutritionists have studied the process by which the body adjusts its activity rate to lower food intake, etc. But this slowing-down of activity has consequences for future output since lack of adequate nourishment influences effort and productivity adversely. This process has been studied in general but not integrated into the economics of famine (Bliss and Stern 1978; Dasgupta and Ray 1986/7). But is is clear that low output this year would lead to low employment and low wages which in turn perpetuate such low output in the following years. Thus droughts in one year can lead to famines in subsequent years, notwithstanding that the rainfall in the subsequent year may have been normal.

A third set of dynamic considerations are in terms of the interactions between different socio-economic groups. The asset holdings of these groups, their saving propensities, and the opportunities for investment open to them in normal times would to a great extent decide how long they can last when faced with shortage. In some sense, adequate reserves of grain and assets confer social power and determine who benefits from famines. After the famine is over, the changed distribution of assets means that those who sold their assets start again in a vulnerable position and those who were prosperous to begin with have strengthened their position. *In extremis* people may move from one group, say pastoralists to another, landless labourers, or may migrate to urban areas. Thus Sen noticed that in the Great Bengal Famine many people tried to get work in rice-harvesting and other such operations though previously they were in the non-agricultural part of the rural economy. Large-scale migration of refugees has been a noticeable feature of the Sahel and the Ethiopian famines.

For those who have assets and stocks of foodgrain, famines may afford opportunities for profits. A perennial controversy about famines is whether they are caused by the hoarding of foodgrains by the farmers or by the grain merchants. Standard economic theory has been somewhat divided on the issue as to whether speculators cause instability or they help markets clear more rapidly and hence prices may be lower than otherwise. The issue concerns the profitability of stockholders' behaviour about hoarding or releasing foodgrains. This brings up the question of the expectations of the traders as to how the prices are going to move and whether such expectations reveal stabilizing influences or not. But the issue concerns not only how traders form expectations but also how consumers and producers form them as well. Only empirical work can give us a classification of markets where stockpiling is stabilizing and where it is not. Ravallion has investigated this question for Bangladesh and found some evidence of destabilizing speculative behaviour (Ravallion 1985).

MODELLING FAMINES

We now have the benefit of a number of descriptive studies of famines, most of them inspired by the entitlement approach. From them a number of common elements emerge as well as significant differences. It is possible to construct a model of the famine – a model in the sense of a set of articulated blocks or systems which interact in specified ways. These systems link with each other through time and display the usual properties of recursivity or simultaneity, exogeneity, or endogeneity. Any such general model must not only be comprehensive in the sense that it must capture the main features of the known famines but it must also leave the possibility open that at the end a famine may or may not occur. Also it must capture the basic premise of the entitlement approach, that none of the usually cited causes for famine – e.g. decline in food availability due to drought, stockpiling, population pressure – is either necessary or sufficient for the famine to occur. Famines are social rather than natural events in this sense.

The basic blocks of this model are:

- The Nature System
- The Socio-Political System
- The Economic System:
    The Food Production System
    The Non-Food Production System (Rural)
    The Non-Food Production System (Urban)
    The Food Delivery System

THE NATURE SYSTEM

Of these systems, the nature system is exogenous with respect to the rest. There is a very long-run connection between the utilization of resources by the economic system which may damage the ecology as in the creation of the US Midwestern dustbowl but for the time perspective within which famines occur, we can take the nature system as exogenous. The nature system will comprise certain stock variables, e.g. the ecology, the land quantity and quality, the other natural resources available. But more important in a temporal context are the flow variables, especially rainfall. It will be the amount, timing, and the geographical incidence of rainfall that will be a crucial exogenous variable. Too much rainfall (floods) or too little (drought) can be a trigger for the chain of events which end in famine. Whether such is the end result or not depends on the conditioning stock variables – the ecology and the structure of the other systems.

Although the nature system is exogenous, it is not necessarily unpredictable. There are regular cycles in rainfall, for example, which can be statistically modelled to gain an early forecast of the likelihood of high or low rainfall. The availability of such a forecast combined with the ability of the socio-political system to respond can avert a trigger such as drought from developing into a famine. (The pioneering work on climatic history is Lamb (1982). See also Wigley *et al.* (1981). For forecasting of droughts see Heathcote (1985).)

A related aspect of the nature system is animal behaviour. Attacks of locusts can cause a collapse of food output even after a normal rainfall, good planting, and fertilizing. The migratory behaviour of locusts is an exogenous but increasingly predictable variable. A different but equally important shock can be the outbreak of disease among the cattle. This can influence both the food and the non-food production systems.

THE SOCIO-POLITICAL SYSTEM

The working of the economic system cannot be divorced from that of the socio-political system. At the macro-level, this involves the nature of the state and of the society. A state hostile to groups within its territory either on racial, ethnic, religious, or class grounds can deliver severe shocks to the availability and distribution of food. The Soviet famine of 1932–3 was most probably a result of the collectivization campaign which led to wholesale destruction of cattle by the peasantry. Civil war is another frequent occurrence which disrupts the production and delivery of food-stuffs. Russia in 1920 and Ethiopia in recent years have witnessed civil wars which seriously disrupted the flow of food as between different regions. Civil wars also illustrate an aspect of the entitlement approach which has not been sufficiently brought out. Entitlements presume a legal framework of property rights; in civil wars there is a dispute precisely about legality. Property owned by someone belonging to one side is seized by the opposite side. Thus entitlements become precarious due to non-economic reasons in civil-war periods.

The socio-political system also tells us about the structure and incidence of taxation. The taxation of agricultural output in one form or another is a central source of revenue for most less-developed countries and even for today's developed countries in their historical period. The method of taxation – direct requisition, income tax, or indirect tax – will determine allocative effects on relative food entitlements. One example of this was given above but more work is needed here.

The socio-political system is also crucial for the administrative/political response to the threat of famine. The Famine Code devised by British civil servants in India during the late nineteenth century has been much studied recently as an example of the practical response based on an implicit entitlement approach (Rangasami 1985). In more recent famines in India the effectiveness of different provincial administrations even within a common famine-relief tradition has been contrasted as between Maharash-tra and Bihar (Dreze (1986) on Maharashtra; Brass (1985) on Bihar). It is not clear what it is that determines the competence of different socio-political systems in responding to threats of famine. Thus in the Sahel and Ethiopia, it is often international rather than domestic famine-relief that seems to be bearing the brunt whereas in Kenya in the 1984/5 famine, it was the domestic administration which was able to avert the famine.

A much deeper aspect of the socio-political system is the social struc-ture–kinship, behavioural norms, inter-group conflicts, religious taboos, etc. These will determine whether the food is shared out equitably and at what level of social aggregation. Certain groups, e.g. single, unmarried, or widowed women, may get systematically marginalized, or allocation norms may discriminate against the old and infirm and concentrate on the survival of the able-bodied who have the best chances of survival. Such micro-variables can only be captured by anthropological fieldwork and economics has little to say on such things.

The socio-political system has to be thought of as exogenous to the economic system. Again if a famine condition becomes endemic, this may have repercussions on the stability of the socio-political system, but for the purposes of understanding famines and devising measures to avert them, it is best to take them as given.

THE ECONOMIC SYSTEM

The economic system will, in general, be endogenous. Variables such as stocks of foodgrains or of capital equipment or of cattle will be given by past decisions, i.e. be lagged endogenous. So will be the given distribution of financial assets and other endowments. The vulnerability of a community to a shock such as failure of rainfall will depend to a large extent on the size and distribution of its reserve stocks of foodgrains as well as assets which can be cashed to buy food.

In normal times again the interaction between the food and the non-food systems is simultaneous. They determine the mutual demand for each other's products and the relative prices. But as we have seen in abnormal periods (in the various senses described in the section above) the food production system becomes pivotal and asymmetric with respect to the non-food system. Thus we look first at the food production system.

The Food Production System

Famines are ultimately about the insufficient access of a large part of the community to food. The food production system is the crucial block. The input-output lag, between sowing and harvesting, is a fundamental feature. The dependence of the system on rainfall as against other sources of water, the cropping pattern, and the technology of cultivation are all important variables. Thus, if the rains fail, the possibility of changing the sowings to another less water-demanding crop will allow the farmers to salvage something out of the situation. The size of crop standing in the field was used as an early indicator of size and harvest in Indian revenue assessment (the so-called annawari system). The vulnerability of the harvest to slight delays in getting the labour or equipment is another factor.

These various factors – dependence on rainfall, possibility of changing the crop, the size of the standing crop as an early indicator of final harvest, the vulnerability of the harvest to last-minute delays – are ordered in time along the input–output lag, i.e. the former occur earlier in the input–

output sequence. The earlier the signal of a possible failure in harvest, the more time there would be for the socio-political system, especially the administrative system, to respond to the signal.

The food production system in normal times determines the output of food, as well as its demand for inputs. Chief among the inputs is hired labour as far as the famine incidence is concerned. If a harvest failure is indicated at an early stage – low sowings, low rainfall – then we know that there would be little demand for hired labour. This would be a sure sign that the labourers' entitlements will shrink unless a relief works programme is immediately initiated. The labourers would otherwise face low employment and low wages. The mode of wage payment, cash or kind, will further determine the size of the price effect. (Some of these points are developed further in Desai (1986b).)

The tenurial relations in food production – sharecropping, tenancy, owner cultivation, co-operative cultivation – as well as the distribution of cultivable land will be the next set of variables to look for. They determine the income of the various agents. Depending on the size of the harvest, the wage bill (in kind) paid out, farmers' own consumption, we can calculate the size of the marketed surplus. The price at which food will be sold to those who have to purchase it as against those who can appropriate it directly will depend on the size of this harvested surplus along with the demand emanating from the non-food production system.

The Non-Food Production System (Rural)

Descriptive studies of famines have revealed a variety of non-food-growing activities which are the second major source of livelihood. Fishermen in Bengal, pastoralists in Sahel, jute growers in Bangladesh – all such activities typify this sector. The point is that the bulk of the poor in LDCs are in rural areas and they are frequently net purchasers of food rather than sellers. Thus, while much development literature talks of the terms of trade between agriculture and industry or between rural and urban areas, it is the intra-rural terms of trade between the food and non-food sectors which matter in times of shortage.

The rural non-food production system can be as much affected by a weather shock – lack of rainfall, for example – as is the food production system. The state of the ecology will matter equally to this system. Thus the nature system has an equal impact on both the rural economic systems. In addition the non-food system may be affected by cattle disease, mercury poisoning of fish, etc. which are specific shocks to this sector. Thus a famine could start because, say, the incidence of foot and mouth disease reduces the incomes of the pastoralists and hence their entitlement to food even without the food supply having suffered.

In normal times, there would be mutual dependence of the food and non-food sectors. They will demand things from each other and sell products as well. To the extent that the non-food system sells inputs to the food system, if activity rates are low in the food sector, then the repercussions will be felt everywhere. If the non-food sector sells consumption goods or services, the impact of a rise in the relative price of food will be adverse. This is because, as I showed above, the share of food expenditure will rise and money spent on other items will fall. Thus during the Great Bengal Famine many who were providers of rural services, e.g. the village barber, suffered from a collapse of the market for the things they had to sell.

The Non-Food Production System (Urban)

This block is mainly a demander of food, and to the extent that the food production system may use sophisticated inputs, e.g. fertilizers, a provider of inputs. In general, it controls credit flows into the food production system. In many famine-prone economies this sector is underdeveloped and hence unable to provide infrastructural support such as transport which would help improve the food delivery system. In famine situations, this system is a demander of food and often has limited quantity of goods to offer in exchange. Such for example was the case in Russia during the Civil War.

But although this sector is a burden on the economy in famine periods, typically there is a grain storage capacity here. Thus the people seeking food and work often migrate from rural areas to urban areas. This is where they are likely to find jobs and also where there is a better chance for food to be available. Urban deaths are also more visible and bring pressure to bear upon relief agencies, national or international.

The Food Delivery System

This system is the network of traders and transporters who store, transport, and sell grain. The merchants who hoard grain are part of this system as is any public distribution system. The ability of this system to cope with famine will depend upon the density of the network. Thus in the Sahel, it has been difficult to organize relief chiefly because the transport network has been thin, lacking roads as well as suitable transport equipment. This can happen if food markets are isolated and the transport of foodgrains is a seasonal activity. In such a case, the transport of food from the coast to the interior becomes problematic. The efficiency of the public relief system, domestic or international, presumes the existence of a dense network.

As already mentioned above, a contentious aspect of the food delivery system is the role of traders, especially their stockholding behaviour. Economics does not support the popular prejudice against the trader/ middleman (often also ethnically different from the local population). By buying at harvest time and selling at lean times, the trader makes a profit but smooths out price fluctuations. The debate then centres around whether, by delaying release of foodgrains from stock on the expectation that prices will rise even further, the trader makes excessive gain. Theoretically, the argument can be made on both sides and can only be settled by empirical investigation. Much more work is necessary on this aspect as on the influence of public distribution systems on price fluctuations.

A common misconception about the entitlement approach to famines is that somehow bringing in food from outside is irrelevant. This is not true. Famines may occur with or without food shortage but increasing food availability is an important plank of famine relief. The other plank, of course, is to initiate an income generation (relief work) programme that will enable people to buy the food.

THE MODEL AND ITS INTERRELATIONS

The interrelations between the different blocks are given in Fig. 3.3. The exogenous blocks – the nature system and the socio-political system – have arrows emanating from them but not directed at them. Beside each arrow the major influencing activity is also given but this is only as a signal for a cluster of variables. Thus 'taxation' is a summary term for a whole gamut of ways in which the state extracts resources from the economic system as well as the benefits it may give. The economic system is endogenous and there is a simultaneous interaction amongst the various blocks as is clear from the chart. Although there are no econometric models available at the present of famines, any future modelling activity will have to start with a system structure such as given in Fig. 3.3. Indeed as details are filled out we shall learn more about the dynamics of the system.

An alternative way to characterize the system would be by examining the relationship between different socio-economic groups within and across the various blocks. Thus the food production system has landlords, farmers, sharecroppers, and landless labourers. Other systems will have similarly employers, self-employed, employee groups. Their interrelations would be structured by the causal flows in the 'objective' system structure but there would also be modifications due to social and political features. Thus there may be ways of putting pressure on the socio-political system from the economic system to alter the burden of taxation on the pace of relief operations. Some of the disaggregated pattern of interrelationships

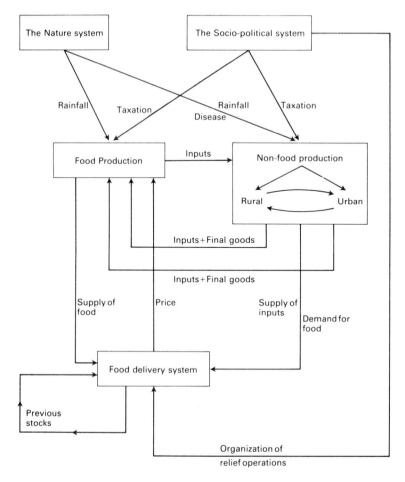

Fig. 3.3. The systems structure of famines.

have been formally covered in Appendix B of Sen's book (1981a) and the actual case studies in his book, and subsequently have filled out the details.

In Fig. 3.4 these interlinkages are described as in normal (non-famine) times. In the food production system we have landlords, tenant farmers, owner cultivator farmers, sharecroppers, and landless labourers. They exchange land for rent (landlord–tenant), labour for wage (landlord–tenant farmer–owner cultivator–labourer), inputs for grain (farmer–sharecropper). Each of the parties except the landless labourer (though unlikely for sharecropper) sells their surplus grain to the trader. The non-food system has been collapsed into three groups, employers, self-employed, and employees. They appear mainly as purchasers of foodgrains. There is no

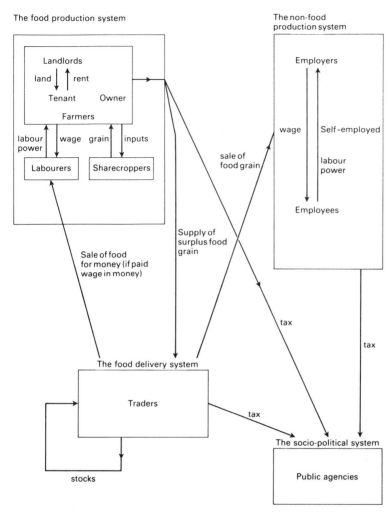

The food production system

The non-food production system

Fig. 3.4. The systems under normal conditions.

attempt at present to specify the relations within the non-food sector in detail but the wage and employment level and the asset endowments will, of course, determine the incidence of starvation when famine occurs.

The essence of the problem then is that many of these regular links are snapped or operate at an altered intensity in famine times. Thus the labour/wage exchange link could snap totally and labourers fall on relief. Similarly the delivery of surplus food to traders could be much reduced and traders may have to rely on previous stocks. The non-food system becomes dependent on the traders and the public relief for their supply of food and also for jobs. To illustrate this reversal of the normal pattern in famine times, Fig. 3.5 describes a different pattern. Thus, there is no delivery of surplus grain from the food production system and there is no employment for the labourers in the food production system or for employees in the non-food production system. Thus the public agencies have to step in to augment the previous stocks of the traders and to provide a relief wage in return for labour services to the workers in the food and non-food sectors. For simplicity the food growers – landlords, farmers, and sharecroppers – are assumed not to be buyers of grain.

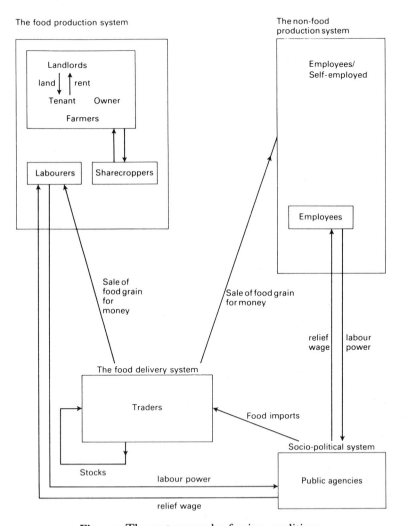

Fig. 3.5. The systems under famine conditions.

Of course, there are simplifications in all of the Figs 3.3–3.5. The relief system may not work as smoothly as the figures assume. Even the food production system may need to buy food. There is scope for further details, for example the allocation within households, migration from the non-food system to the food system of workers, etc. The purpose of these figures is, however, to convey the central features of the system as emphasized by the entitlement approach.

CONCLUSION

Famines are a relatively new topic of study for economists after a lapse of a hundred years or more. They have forced economists to examine the economy as more than merely exchange of goods and services by having to bring in social and political interrelationships and by exploring the nature of the economic system in a state of malfunction. The entitlement theory has proved fruitful precisely because it offered a simple but flexible framework for integrating the non-economic with the economic. But even here it has to be admitted that our understanding of famines is still very much at the beginning not the end of the process.

LIST OF REFERENCES

Bliss, C. J. and Stern, N. H. (1978). Productivity, wages and nutrition, part I, theory and part II, some observations. *Journal of Development Economics*, **5**, 331–62, 363–98.

Dasgupta, P. and Ray, D. (1986/7). Inequality as a determinant of malnutrition and unemployment. *Economic Journal*, December–March, 177–88.

Desai, M. (1972). The role of exchange and market relationships in the economics of the transition period: Lenin on the tax in kind. *Indian Economic Review*, 7, 61–8.

Desai, M. (1984). A general theory of poverty? (review article on Sen (1981).) *Indian Economic Review*, **19**, 157–69.

Desai, M. (1986a). Rice and fish: a general equilibrium approach to entitlement failures. (LSE, unpublished discussion paper.)

Desai, M. (1986b). Famine anticipations and early warnings. (Paper presented at United Nations World Institute for Development Economic Research Conference on Hunger and Poverty, Helsinki, July.)

Dreze, J. (1987). Famine Prevention in India. In Sen and Dreze (1987).

Greenough, P. (1982). *Prosperity and misery in modern Bengal*. Oxford University Press, Oxford.

Heathcote, R. L. (1985). Extreme event analysis. In Kates, Ausubel, and Berberian (1985), pp. 369–402.

Jasnyi, N. (1972). *Soviet economists of the 1920s*. Cambridge University Press, Cambridge.

Kates, R. W., Ausubel, J. A., and Berberian, M. (1985). *Climate impact assessment*. Wiley, New York.

Koopmans, T. C. (1957). *Three essays on the state of the economic science*. McGraw-Hill, New York.

Lamb, H. H. (1982). *Climate, history and the modern world*. Methuen, London.

Lenin, V. I. (1971). The tax in kind. *Selected works*, Vol. 31. Progressive Publishers, Moscow.

Mitra, A. (1985). The Meaning of Meaning (review of Sen (1981)). *Economic and Political Weekly*, **17,** 27 March.

Rangasami, A. (1985). Failure of exchange entitlement theory of famine: a response. *Economic and Political Weekly*, **20,** 12 and 19 October.

Ravallion, M. (1985). The performance of rice markets in Bangladesh during the 1974 famine. *Economic Journal*, March, 15–19.

Sen, A. K. (1977). Starvation and exchange entitlements: a general approach and its application to the Great Bengal Famine. *Cambridge Journal of Economics*, **1,** (1).

Sen, A. K. (1981a). *Poverty and famine: an essay on entitlement and deprivation*. Clarendon Press, Oxford.

Sen, A. K. (1981b). Ingredients of famine analysis: availability and entitlement. *Quarterly Journal of Economics*, **95.**

Sen, A. K. (1986). Food, economics and entitlements. Lloyds Bank Review, April.

Sen, A. K. and Dreze, J. (1987). Hunger and poverty: the poorest billion. Proceedings of WIDER Conference on Food Strategies Research.

Sorokin, P. A. (1975). *Hunger as a factor in human affairs* (ed. T. Lynn Smith). University of Florida Press, Gainesville, Florida.

Srivinasan, T. N. (1983). Review of Sen (1981a) in the *American Journal of Agricultural Economics*, February.

Thompson, E. P. (1978). *The poverty of theory*. Merlin Press, London.

Wigley, T. M., Ingram, M. J., and Farmer, G. (1981). *Climate and history: studies in past climates and their impact on man*. Cambridge University Press, Cambridge.

4

THE ECOLOGY OF FAMINE

P. J. Stewart

INTRODUCTION

Do people get the famines they deserve, or is famine inflicted by the strong on the weak? Starkly put, these are the two extreme positions between which views run. The first sees the event as essentially local, the consequence of an imbalance between people and their environment, while the second looks on it as a manifestation of distorted national and global patterns of distribution. The first lays the blame on the ignorance or heedlessness of the afflicted population, the second on the greed and insensitivity of those who dominate the world economy.

The theory that famine is locally self-inflicted has been clearly stated, for example, by Hardin (1986). Claiming to speak in the name of an objective human ecology, he assumes that any given territory has a physically determined carrying capacity; if its inhabitants multiply their population beyond this limit, their transgression of natural law is relentlessly punished by famine. Other countries, he says, should not intervene, for that will only prolong the disequilibrium and worsen the damage.

There are three fallacies in the belief that people get the famines that they deserve. Firstly, it assumes a fixed biological carrying capacity, and takes no account of the fact that the number of people who can be fed by a given piece of land depends partly on their techniques of food production, on the diet that their culture has taught them to accept as normal, and even on the physique and metabolism that they have developed, whether as a consequence of their dietary history or of their genetic inheritance. Even if food production falls in a bad year, carrying capacity may, up to a point, be maintained by substituting more efficiently for less efficiently produced foods, for example plant protein for animal protein.

Secondly, the theory makes the Malthusian assumption that excessive population growth is the primary cause of impoverishment, and that the cure for poverty is therefore a decline in numbers. The evidence is, however, that the causation is in the opposite direction: that impoverishment (which may have many causes) is one of the factors that can cause people to produce too many children, because each individual fears that too few will survive to adulthood. Rising prosperity appears to have been the primary factor bringing about a fall in birth rates. As for the causal relationship with impoverishment, population growth has in many cases stimulated a faster growth in productivity and income, leading eventually to the conditions for its own cessation. Indeed, many economists believe that the declining population now seen in some of the richer countries is a factor causing national income to stagnate or decrease. All of these phenomena can be observed in European history over the past two centuries.

Thirdly, the theory that famine is deserved confuses carrying capacity with self-sufficiency in food production. This leaves entirely out of account the ability of societies to engage in trade, exporting some goods and services to pay for the import of others. Even an agricultural district may depend on outside supplies of essential foods, paying for them by producing a surplus of other foods. In fact all countries import food, and many are net importers. If a country or region is unable to import sufficient food to complement its own production, this is because of factors such as markets, exchange rates, and the history of investment. Economics, in other words, is part of human ecology. Economic carrying capacity is not the same as biological carrying capacity. There is, thus, no environmental determinism in famine. Even a desert can support a city, provided its inhabitants have something to sell to the outside world, as the citizens of Mecca, for example, have proved over more than fourteen centuries. Conversely, there is no such thing as an environment proof against famine. Even the fertile farmland of Europe has at times allowed its people to starve, and economic breakdown could conceivably permit this to happen again.

There is, however, a sense in which famine is, if not deserved, at least to some extent self-inflicted. Over-cropping and over-grazing have reduced vast areas of once fertile land to an eroded state in which famine is, if not inevitable, at least less improbable than need be. Declining productivity has annihilated the surplus of once prosperous farming or grazing communities, converting their economy to one of fragile subsistence. Reluctance to emigrate has prevented the adjustment of population to changing conditions. Failure to invest in, or to attract investment in, alternative forms of production has left the community dependent on outside charity in times of hardship.

Subsistence modes of existence existed long before the development of the exchange economy, and in many cases they were in stable equilibrium with the environment, removing nutrients at a rate which did not exceed that at which they were restored. However, sooner or later, most such cultures have eroded their natural resource base, in which case their situation is now much the same as that of formerly prosperous rural communities that have withdrawn from the exchange economy because of failing surpluses.

This chapter will examine the ecological processes that can lay a people open to famine, and will suggest ways in which they can be halted or reversed. Three stages may be distinguished. The first is the preparation of famine, in which over a long period there are signs of declining productivity. The second is the crisis of famine, during which certain events ease the pressure on the system while others may store up fresh troubles for the future. The third is the aftermath of famine, which may be made into the opportunity to prevent a recurrence or may simply prove to be the preparation for the next famine.

THE PREPARATION OF FAMINE

THE POINT OF DEPARTURE

A good case can be made for seeing drought as an immediate cause in all famines (Dar 1983). It would be simplistic, however, to conclude that famine has its roots in climate. The term climate in fact covers two quite different sets of phenomena. It is useful to distinguish the background climate, which is characterized by the pattern of inputs of solar energy and precipitation, from the effective climate, which is characterized by the pattern of atmospheric temperature and humidity. To a given background climate corresponds a whole range of possible effective climates, the differences between which are caused essentially by differences in the nature of the surface and its capacity to absorb incoming energy and rain-water. Modification of the effective climate over very large areas can alter the background climate through its influence on cloud formation and on atmospheric dust.

It is life that determines the effective climate, through its capacity to reverse the general trend of physical systems towards increasing disorder. Where, in the beginning, sun, wind, and rain beat irregularly on bare rock, living things build a fabric of soil and vegetation in which conditions are relatively constant. As long as there are spaces in the ecosystem and adapted species to fill them, the mass and complexity of the system increase until, in what is called the climax, they are limited by the physical inability of plants, in the presence of the given assortment of plant-eating organisms and their predators, to stand taller or to reach deeper with their roots.

Climax ecosystems offer the creatures that inhabit them the greatest stability possible, given the conditions of background climate, geology, and topography. Their soil stores a reserve of water, most of which would otherwise have run off at a rate dictated by the slope, and much of which would have found its way to the sea. Evaporation of this water *in situ* uses up large amounts of the sun's energy, which would otherwise have raised the temperature of the ground and of the air lying over it. Cooler air can take up less water vapour before being saturated, so that evaporation is slowed down. Vegetation affects evaporation also by reducing wind speed and air turbulence.

The effect on growth is threefold: there is more water available in the soil; plants lose water at a slower rate, and can continue photosynthesis longer; and at lower temperatures they also use up in respiration less of what they manufacture. A climax ecosystem thus optimizes the effective climate about itself, prolonging growth into what otherwise would be part of a period of drought, and increasing the amount of net production that can be accomplished with a given amount of water. The more intense the sunlight, and the more irregular the rains, the more important this effect. Its greatest impact is under the climates with marked dry seasons that predominate in the tropical and mediterranean zones.

Furthermore, vegetation protects the soil on which it depends for its reservoir of water. Living foliage and stems shade the ground from direct sun, and absorb much of the mechanical energy of falling raindrops. Discarded foliage forms a mat, covering the surface and replenishing the soil organic matter that is an important source of nutrients. Roots bind the soil together, and, when they decay, they leave channels along which gases are easily exchanged between soil and air. Dead and living plant materials feed the myriads of organisms that live in the soil, and which constantly renew its structure and chemical composition, converting organic matter into soluble nutrients that plants can recycle.

Perennial plants in general, and trees in particular, play the most important part in assuring the stability of living communities. Their size enables them to affect a greater volume of soil and air than annual plants, and it allows them to hold large reserves of food within their own bulk. With their deep roots they draw up water and nutrients that would be inaccessible to annual plants. They cover the ground in all seasons, and their long lifespan, often of hundreds or even thousands of years, makes for systems with remarkable constancy.

These effects of recycling and stabilization are again most important under tropical and mediterranean climates. There the prevailing high temperatures make for rapid breakdown of dead material into easily soluble inorganic chemicals, and, unless these are rapidly taken up by plants, they are soon washed out of the soil and lost to the ecosystem. Certain combinations of geology and hot climate allow compounds of iron, aluminium, or calcium in an unprotected soil to be precipitated out as hard pans. Where these form on the surface or are exposed by erosion, they are almost completely sterile. Only the presence of a mass of vegetation can here guard against degradation into virtual desert.

Vegetation forms a reserve of food for plant-eaters, and this can be run down in adverse conditions and allowed to build up again in favourable times. Most of the animals in a climax community can continue to live normally even in a year of poor plant growth. The degree of security afforded in this way depends on the mass of plant matter per unit area, which in turn depends on the structure of the animal community, and on the history of the territory. Where there are many species of grazing and browsing animals, some of them too large to be effectively controlled by the available carnivores, or where rainfall is very slight, trees may be unable to close into a forest, and the climax formation is grassland, with a relatively light plant mass.

Outside the equatorial rain-forests, where conditions have remained almost constant for millions of years, the ecological climax is an abstraction rather than a reality. The climatic changes of the past million years have repeatedly destroyed formations of living things and caused their migration to and fro across the continents. By the time the last ice age ended, human beings were already present on all the continents. Our species therefore started its history amid living systems that were generally less robust than theoretically possible, and our presence must in many places have prevented the movement back towards climax conditions from going very far.

Nevertheless our hunter–gatherer ancestors were able to colonize ecosystems that were in general far richer than we see today, in both mass and variety. They were able to exploit the produce of a natural capital whose creation had cost them no effort. Living with simple ideas of their needs, theirs has been called 'the original affluent society' (Sahlins 1974). Some idea of the abundance of natural produce in sparsely inhabited climax communities is given by the early settlers' description of New England (Cronon 1983). Even the hunter–gatherers of today, surviving in the marginal environments rejected by agriculture, appear able to satisfy all their needs as long as they are left without interference.

DEGRADATION

Over-cropping

The movement down from climax conditions – or from the nearest approach to them – may well have begun long before the invention of agriculture. Human beings have used fire for hundreds of thousands of years, and the deliberate setting of bush fires to flush out game is recorded for many peoples. Later, the invention of the bow and arrow gave the power to eliminate rival predators, and this may have affected the balance with herbivores. Both these factors must have favoured the expansion of grassland at the expense of forest. Analysis of this history is difficult because human colonization of the globe coincided with the climatic upheavals of the late pleistocene.

The early history of agriculture is almost equally obscure. It may have begun with unsystematic weeding around favoured plants, and with the observation of growth on old campsites where seeds had been discarded. This would then have been systematized into deliberate sowing and tending on favoured sites within a territory, which in turn would have developed into shifting cultivation, with people returning regularly to old sites after a period of bush fallow, and becoming mainly dependent on their produce. Such methods are today used by an estimated 250 million people in tropical Asia, Africa, and Latin America (FAO 1978).

The bush fallow of shifting cultivation consists normally of forest vegetation. If it is allowed to grow long enough, this re-establishes conditions more or less close to those of the climax, providing the farmers with good growing conditions for their next round of crops. Cleared areas represent a brief phase in a long cycle and therefore occupy a small fraction of the total area, forming islands in a sea of forest and enjoying the climate generated by this mass of surrounding vegetation. In some places shifting cultivation appears to have continued for thousands of years without serious damage to soil or plant life. In the present century, however, the increase in human numbers has led to a shortening of the fallow period, preventing full recovery of the system. Migration into hitherto unfarmed forest has brought this degradation to new areas, making it now into one of the principal causes of deforestation.

The transition from shifting to settled agriculture has not everywhere badly damaged the soil inherited from natural ecosystems. Restoration of chemical nutrients has been achieved by the spreading of manure, compost or fertilizer, or by the adoption of crop rotations including a nitrogen-fixing leguminous element. Protection against mechanical erosion has been achieved using techniques that cover the surface during sensitive periods, and which slow down and canalize the flow of water. However, these methods are fully effective only where the forces of erosion are relatively mild, which for the most part means in the temperate zone.

In the tropics, conservationist farming has used techniques that maintain some vegetative cover throughout the year. In practice this has meant mixing annual crops with perennial ones, or with trees retained from the original forest. Large-scale cultivation of purely annual crops has been possible as a continuing system only where shallow slopes have made for slow erosion, or where a very deep, loosely structured, and mineral-rich rock has constantly replaced the soil lost, for example, on the island of

Java, or where a great river has replenished its flood plain and delta, for example the Mekong.

Over the past couple of centuries there has been a growing movement of people with temperate notions of farming into the tropics, at first as settlers and more recently as technical advisers and as purveyors of agricultural machinery and chemicals. Generations of students from the tropics have been educated in temperate schools of agriculture or their tropical offshoots. This has resulted in disastrous attempts to 'modernize' tropical farming methods by eliminating perennial plants and sowing annual crops in large, mechanically prepared fields. An example was the Brazilian attempt to move landless peasants from the Nordeste, and to settle them on new farms claimed from the Amazonian rainforest. The programme had to be abandoned after a few years because of the rapid loss of soil fertility.

Another reason for the speeding-up of erosion has been the expansion of cultivation onto increasingly unsuitable land, under the pressure of rising populations. Most of the best sites having already been occupied by farming, people have cleared the natural vegetation from steeper and steeper slopes, whereas many experts believe that permanent cultivation is safely practised in the tropics only where the slope is less than 1 per cent. Similarly, farming has been pushed on to thinner and poorer soils, where a larger area has to be used for a given amount produced, and where less intensive and careful techniques are therefore likely to be used.

Over-grazing

Greater still than cultivation in its effect on soil and vegetation has been pastoralism – the extensive grazing of animals as the main source of livelihood. This probably has a double origin, arising in some cases out of the domestication of animals by farmers, and in others out of the following and progressive control of wild herds by hunter–gatherers. Evidence of the former origin is to be found in the presence of herbivores in the archaeological remains of neolithic farming communities in the Near East. A process of the latter kind is to be seen in the traditional way of life of the Lapps.

The work of modifying soil and vegetation under pastoralism is carried out mainly by animals instead of people, and a small human population can control a large number of herbivores, with several beasts grazing for each person. Mixing animals of different species spreads the effect over the full range of plant types, with goats, donkeys, or camels attacking woody perennials, while sheep and cattle feed on annuals or soft perennials. The use of fire further magnifies the impact, clearing large areas of denser vegetation, to be replaced by tender regrowth. Under such pressure, grasses become dominant; unlike those of most other plants, their shoots grow from the base, and can tolerate repeated biting-off of their tips.

Ecological adaptation to the animals includes not only that of the vegetation, but also that of the people, who learn to live on a diet very high in animal protein and fat. Most significant is the ability, found in most pastoral peoples and in others derived from such origins, to digest milk throughout life, the enzyme for which (lactase) is normally absent from mammals after weaning. It is not clear how far this peculiarity is genetically transmitted and how far it is acquired by childhood training, but the selective pressure in its favour is enormous, for fatal gastroenteritis sooner or later claims those who drink milk without having the necessary enzyme (McCracken 1971; Kretchmer 1972).

Adaptation to a diet high in animal protein makes human beings very costly in ecological terms. About four-fifths of the energy content of fodder is used in the metabolism of the grazing animals – much of it simply in keeping them warm and moving them about. It thus takes typically five times as great an area of land to feed a population of grazers as to provide for an equivalent number of people eating mainly plant foods. The ratio becomes even worse if the productivity of the land is reduced by over-grazing.

The tendency to over-grazing is generally greater than that to over-cropping, because the mobility of herds leads to separate ownership of them and of the land: the animals are usually private property, whereas the land is public. An individual's gain from adding to stock is greater than that person's share in the collective loss of productivity, and so each owner is rewarded for acting against the public interest. Degradation being a slow process, it does not generally create problems acute enough to bring a change to intensive management and control.

Over-grazing is similar to over-cropping in its effects on soil and vegetation. The trampling of the ground by hooves adds to the danger of erosion, compacting the surface, reducing its permeability to water and gases, stifling biological activity in deeper layers, and increasing run-off and erosion by rainwater. Such effects are only partly compensated by the contribution of animals to the cycling of nitrogen in the system. Erosion by wind is relatively important, simply because grazing can extend to drier lands than can rain-fed cultivation.

Fully developed pastoral cultures are known from the earliest historical records, following large mixed herds over vast territories. The speakers of the original Indo-European language had a vocabulary that suggests such a way of life, and their supposed place of origin in the post-glacial grasslands of Eurasia has since been occupied by a succession of other pastoral peoples – most notably by the Huns, Magyars, Turks, and Mongols. The ancient Near East saw repeated migrations of pastoral peoples such as the Hyksos into Egypt, the Hebrews and Aramaeans into the Levant, the Akkadians and Assyrians into Mesopotamia, and finally the Arabs over the whole area.

Northern Africa has shared in the history of the Near East, and even as far south as Ethiopia the dominant languages derive from an ancient colonization from South Arabia. Over much of the rest of sub-Saharan Africa the presence of the tsetse fly has inhibited the spread of pastoralism, but there are very important grazier peoples, most notably the Masai in East Africa and the Fulani in West. In the Americas, it seems likely that the rulers of the Inca Empire, with their huge flocks of llamas, were pastoral in origin, and the Aztec taste for human flesh may have been a transformation of pastoral meat-eating.

The pastoral peoples have a place in world history quite out of proportion to their numbers or their level of civilization. Travelling light, driving their food supply along with them, and early mastering the use of animals in warfare, nomadic graziers made highly effective invaders, as the Arab thinker Ibn Khaldun pointed out long ago. They easily overwhelmed farming peoples, the bulk of whose population was tied to work in the fields, and whose armies depended on bulky food-stocks needing expensive transport. It is only in recent centuries that sedentary peoples have come to live in security from nomadic graziers.

Spreading out over Eurasia only a few millennia after the end of the Ice Age, the herds of the grazing peoples halted or reversed the development towards climax vegetation. Their subsequent invasion of farming civilizations brought pastoral habits of diet and land use to large areas that would otherwise have borne the relatively light load of cultivation. Particularly important was the introduction of the milk-drinking habit into the civilizations of Europe, the Near East, and India, (but not those of China, Japan, or South-East Asia), from which it subsequently spread to most of the rest of the world.

Deforestation

Forests are destroyed not only to make room for grazing or cropping, but also by over-exploitation of their timber and firewood. Until the modern development of international trade this was a danger that existed only near to large centres of population, or in areas with difficult climates. Most cultures had no need of rules to regulate their tree-cutting to ensure regeneration. In recent centuries, however, local demand for wood has greatly increased as populations have increased. At the same time, commercial logging for world markets has come to be an equally important use.

Logging by itself has not generally destroyed forests in the tropics. In Latin America and Africa it was rare to find more than a few commercially valuable trees per hectare. Exploitation damaged the remaining forest but did not prevent it from growing again, though with a different combination of species. It is only in South-East Asia that it has been profitable to clear-fell large areas. Even this did not necessarily lead to loss of the forest, except where topography or soil were particularly difficult. It is rather by opening up hitherto inaccessible land to cropping or grazing that commercial logging has been destructive.

Demand for firewood, on the other hand, has long been a major cause of deforestation near to cities and in densely peopled rural areas. Further from human settlement, charcoal-burning has had a similar effect. Over-exploitation has often been favoured by the same factor as on grazing land: individual exploitation of a collectively owned resource.

Farming

The English word 'farmer' means someone who practises either cropping or grazing or some combination of the two. For the purposes of this chapter it will be useful to distinguish four modes of production: that of the grazier, that of the cropper–grazier, who depends mainly on the products of grazing, that of the grazier–cropper, who depends mainly on those of cropping, and that of the cropper, who may keep animals, but who stables them and feeds them on crop products. The latter three can be referred to collectively as 'farmers'. Where there is no need to distinguish between cropper–graziers and grazier–croppers, they will be jointly referred to as 'mixed farmers'.

Transitions between these four modes take place with varying degrees of difficulty. Graziers rarely become farmers, lacking the complex skills and shunning the back-straining work of cultivation. Mixed farmers, on the other hand, seem to turn easily into graziers, as in the American 'Wild West', and in the plains of Argentina and Australia. Under recent pressures, many mixed farmers in Europe have become croppers, plough-ing up their pastures, and feeding their animals indoors.

Properly managed, mixed farming can remedy some of the erosive tendencies of cropping and grazing. Pasture can be used as a fallow between periods of cropping, and cultivation loosens the soil compacted by trampling hooves. The manure of grazing animals improves the nitrogen cycle, without the laborious spreading that is needed with confined animals. On the other hand, bad mixed farming can be more damaging than cropping, with different erosive processes complementing each other, and with animals destroying what is left of the woody vegetation.

Since grazing is more erosive and more extensive than cropping, the various types of farming can be ranked in order of potential destruction according to the importance of the grazing component. This differs not only for climatic reasons, being strongest in relatively dry countries, where rain-fed cultivation is limited in extent. There is also a strong cultural influence, certain societies having a much greater demand for animals than others.

Culture, Diet, and Land Use

Of the major civilizations, Islam is the one with the strongest need for grazing land. The sheep, with the goat as best alternative, is the preferred animal for the pilgrimage sacrifice, which occurs every 354 days and is considered obligatory for every Muslim household (Stewart 1979), as well as for other celebrations such as weddings and circumcisions. It is only in tropical Asia, where the climate hampers the keeping of sheep and goats, and in the Soviet Union, where, under official pressure, the sacrifice has been largely replaced by almsgiving, that the pressure for extensive grazing is less powerful. The prohibition on pig-meat further narrows the options in animal husbandry.

European civilization, with a diet that traditionally includes plenty of meat, milk, and milk products, is likewise demanding in grazing land. There is not, however, the strong preference for particular animals nor the obligation to kill so many of them at one time that make pasture management so difficult in Muslim countries. Nor has grazing been a cause of severe erosion under the temperate climates of most of the countries that have a culture of European origin. It is above all in tropical America that this demand for grazing land has been destructive.

The indigenous farming cultures of Africa appear to have given a relatively modest place to grazing, and its pastoral peoples were localized. More recently the influences of Islam and the West have greatly increased the demand for meat, and large areas are needed to supply the markets of populous modern cities like Lagos, Nairobi, and Addis Ababa. Western influence has introduced the consumption of milk and milk products to new areas, and the inclusion of dried milk in famine relief provisions may spread tolerance of lactose and thus paradoxically make it harder to find enough land to supply future generations.

Hindu civilization inclines towards vegetarianism, but it has exerted almost the same demand for grazing land as a beef-eating civilization would have done, thanks to the sacred status of the cow. Buddhism is the most nearly vegetarian of the great religions, though practice varies considerably from one country to another. In China, for example, meat-eating was little affected by Buddhist influence, whereas in Japan seafood became almost the only source of animal protein. Chinese animal husbandry, however, is based on backyard stock, mainly pigs and poultry, and on the water buffalo, not on grazing animals. In both countries, as in most of East Asia, milk consumption is absent. There is thus hardly any grazing pressure, and the use of land is relatively parsimonious.

Interactions

Not only has over-exploitation reduced the productivity of individual areas of cropland, grazing land, or forest; in addition, interactions with events on neighbouring stretches of land have reinforced the effects. Harmful influences have also been exported to the territory of communities whose own practices would not necessarily have led to their impoverishment.

The cumulative effect of prolonged degradation has been a shift in the balance between different forms of land use. Former forest has become cropland, former cropland has become pasture, and former pasture has become desert. As the area of cropland has had to be maintained or increased, the net effect has been an expansion of desert and pasture at the expense of forest.

Deforestation has been destructive not only to the land on which the forests stood but also to neighbouring land. The influence on atmospheric temperature and humidity can obviously not be confined by any boundary, nor can the down-slope, nor down-wind effects of erosion. Increasing silt loads and irregularity of flow in the great rivers endanger dams and irrigation systems and add to the risks of flooding at great distances from the mountains where they have their origin. Dust-laden winds and sand-storms can reduce plant productivity and feed systems of sand dunes over wide areas.

The destruction of forests can be a direct cause of malnutrition and potential famine in three main ways. Firstly, forests are themselves sources of a great variety of foods, including wild fruit or nuts, edible fungi, honey, and wildlife. Even animals that live mainly in open country are in many cases dependent on tree cover for part of their life cycle. The effect of forests on the water regime is also important for its influence on streams and bodies of water, with their potential for fisheries.

Secondly, the loss of fuelwood sources reduces the capacity to cook, which is necessary for the proper digestion of many of the foods produced by agriculture, or for the elimination from them of parasites and disease organisms. Longer journeys for firewood also mean increased energy expenditure and nutritional requirements for those who fetch it; it is commonplace for a day's walking to be needed for the collection of enough wood to last only three or four days.

Thirdly, faced with fuelwood shortages, people burn dung and crop residues, which are thus no longer available to maintain soil fertility. The use of dung as fuel depends on its physical characteristics; large cakes, like those of cattle, are easy to collect and use, whereas small pellets, like those of the sheep or goat, are more likely to be left on the ground. Cultural factors that influence the mix of grazing animals therefore also affect the amount of manure that is burnt. India, with its many cattle, suffers particularly in this way.

THE CONDITIONS FOR FAMINE

If not corrected, the various processes discussed above sooner or later reduce soil and vegetation to a minimum. The ecological conditions are then as far as possible from those of the climax, without adequate biological reserves or protection, and closely following the fluctuations of the weather, and of the background climate.

Without a deep soil to soak it up, rainwater quickly runs off and is lost to the locality. What remains is soon evaporated, and the sun's energy is thereafter used to raise ground and air temperatures, lowering relative humidity and increasing the stress on plants, animals, and people. A given input of rain and sun thus produces the worst possible effective climate.

The loss of evaporation and the rise in surface temperatures may in turn cause a reduction of cloud formation and rainfall in a vicious circle that further worsens conditions (Charney et al. 1975; Ottermann 1974). Much of the expansion of the world's deserts can probably be attributed to human mismanagement of soil and vegetation. Worsening effective climate causes the abandonment of much cropland and its transfer to extensive grazing, hastening the process of degradation.

The shrinking reserve of biomass thus comes to include a high ratio of animal to plant material. It is all the more fragile in this form, since livestock has to be fed every day and watered at frequent intervals, unlike the plants adapted to such conditions, which are able to survive long periods of dormancy. The burden of ministering to the needs of their animals causes people to divert precious water resources away from crops, and to sacrifice any remaining perennial vegetation.

A system so reduced has very little resilience. A spell of drought or an attack of animal disease is enough to produce a disastrous loss of productivity. The migration of hungry people and livestock into neighbouring areas overloads their capacity in turn and spreads the danger of famine.

Steep fluctuations in productivity would not lead to catastrophe if the human population were reduced by progressive emigration to the level that can make a living in bad years. This is what probably happened until relatively recently with nomadic graziers, as long as there was vacant or conquerable territory for them to move to. Indeed, periods of drought are seen in some theories of history as the cause of the great episodes of pastoral empire-building.

With sedentary peoples, the scope for a slow and orderly loss of population is far smaller, and modern methods for controlling nomads have produced a similar restriction even for them. The reduction in mortality rates over the past 100 years has caused rural populations to rise instead of falling, further increasing the danger of disaster in bad years.

The conclusion to be drawn is that the over-exploitation and erosion of biological resources create the conditions in which famine can most easily occur. However, it is not suggested that famine is inevitable under such circumstances; to suppose that would be to put the blame on the past and to deny present human responsibility. Governments can very well help to tide societies over short-term difficulties while encouraging long-term movements of population or improvements in land husbandry. It is the combination of erosion with administrative breakdown that is lethal.

FAMINE AND ITS AFTERMATH

THE CRISIS OF FAMINE

When famine comes, it may in itself be a major ecological event. Leaving aside human considerations and looking just at physical aspects, it is clear that the most important consequence is a reduction in the human and animal power available for continuing over-exploitation. This may mean improved growth of natural vegetation when better climatic conditions return.

The extent of any improvement in growth depends on the degree to which the population is reduced and on the nature of their exploitation of the land. Farming communities are the most likely to suffer a steep decline in population, because of their relative immobility, and abandoned fields may quite quickly revert to natural vegetation.

Nomadic graziers, on the other hand, may manage to keep up their human population by taking advantage of their mobility. A reduction in their livestock may be insufficient to make much difference to pasture; under conditions of over-stocking, individual animals eat considerably less than their capacity, and fewer animals will eat the same amount. There may even be an adverse affect if the loss of more selective feeders leads to an increase in the proportion of versatile feeders such as goats, camels, and donkeys. The return to desolation can be quite rapid.

Reduced human numbers may have detrimental as well as beneficial effects. The search for substitute foods can lead to the exploitation of resources that normally would be spared, such as the bark of trees and the tubers of wild perennial plants. Outside the famine area, the outflow of population may have devastating effects on neighbouring regions, which suddenly find themselves in their turn over-populated and subject to degradation.

AFTERMATH

Famine, unlike the chronic undernourishment that afflicts so many of the world's poor, is an event. As such it draws the attention of governments. There is pressure to be seen to be taking measures to ensure an early improvement of the situation. The danger is that these will be hastily decided on the basis of inadequate study, and that some of them will actually worsen the destruction of soil and vegetation.

The actions that may have damaging effects are those that make possible more intensive exploitation of the kind already practised. The introduction of agricultural machinery makes possible deeper ploughing, with added risks of erosion. The provision of new watering points, of veterinary services, and of supplementary feedstuffs allows livestock numbers to build up again to harmful levels. The Sahel famines, for example, have been partly attributed to the boring of Artesian wells after the 1973 famine; these made it possible for graziers to settle in areas that they had previously visited only seasonally, where the vegetation could not support year-round grazing (Sinclair and Fryxell 1985).

WHAT FAMINES ARE WE PREPARING FOR THE FUTURE?

The famines of the future are already being prepared by the continuation of the processes that led to those of the past. As long as over-cropping, over-grazing, and deforestation are allowed to go on destroying soil over huge areas, the prosperity of many rural communities will continue to be threatened. According to a recent joint study, carried out for the UN Fund for Population Activities by the Food and Agricultural Organization and the International Institute for Applied Systems Analysis, erosion threatens to remove from cultivation 17·7 per cent of the rain-fed cropland in the hotter countries, and to reduce the productivity of the remaining land by 28·9 per cent (FAO 1982). The excess of population over carrying capacity, assuming the need for self-sufficiency in food production, is expected to increase from 278 million in 1975 to 503 million in the year 2000, if present levels of investment in agriculture and soil conservation are maintained. The risks of famine are thus increasing.

Even the means to provide emergency relief for famine seem unlikely to improve. The world population is expected to double by the middle of the next century. The average productivity of grazing land and fisheries will probably not increase, and may even decline, so additional food will have to come from arable land. The area of this, however, is not expected, on prudent estimates, to show much sustainable long-term increase over its present total (Revelle 1976). Simply to maintain present average nutrition levels, it will therefore be necessary to double the productivity of farmland, which is likely to be achieved only with difficulty. Unless there are major changes in the technology of food production, the total amount per head of population is thus likely to change little, if at all. If disparities of wealth continue to allow rich countries to feed a large proportion of the world's grain to animals, the amount of food available for famine relief will be limited.

THE PREPARATION OF PROSPERITY

GENERAL PRINCIPLES

The livelihood of most rural communities depends mainly on biological production, which should provide a sufficient surplus to pay for goods and services bought in from outside, and sufficient reserves to tide over hard times. The income per head depends, of course, on costs and prices and on the size of the human population, but, other things being equal, it will be maximized when biological productivity is at its highest level. Since the ecological climax provides the best possible environment for plant growth, the founding principle of rural land management should be to restore and maintain conditions as close as possible to those of the climax.

The essential environmental features of the ecological climax are maximum soil depth and water retention, recycling of nutrients, and protection against the excessive impact of sun, rain, and wind. These combine to moderate the local climate and shorten the periods unfavourable to growth. The more extreme the climate, the more significant the effects. The protective function is relatively unimportant under temperate climate but is dominant in hotter countries. Humid tropical climates also need rapid nutrient recycling, as dead organic material is soon broken down at high temperatures and quickly leached out by abundant rain. Only the constant presence of active vegetation with a developed root-system can ensure that nutrient losses are minimized.

LAND HUSBANDRY

Under temperate climates, it is generally possible to make an approach to climax soil conditions by cultivating and fertilizing between rounds of annual crops. In hotter countries, it is advisable to provide a protective cover of vegetation all the year round; where annual crops are grown, there should be sheltering trees or shrubs, either in intimate mixture or in close juxtaposition.

The fashionable term for systems that combine annual and perennial crops, with or without animal husbandry, is 'agro-forestry'. There are many objections to this expression (Stewart 1981), not least the fact that to the uninitiated it sounds as though it should mean a kind of forestry. In particular the term separates out the area of overlap between farming and forestry, implying that it is a special and self-contained field of study and action, the province of a new set of professionals – the 'agro-foresters' – with potential for conflict and misunderstanding across two boundaries where before there was only one.

Rather than a term that subdivides rural activities still further, what is needed is one that includes them all, with temperate-style agriculture and forestry simply as extreme cases. 'Land husbandry' has been suggested as such an expression, and it will be adopted here. It implies that the various kinds of land use form a continuum and are to be applied separately or in combination to provide appropriate treatment for every possible situation. It follows that the separation of agriculture and forestry – in administration, ownership, education, and research – is a source of confused thinking and ineffective action. Even in Europe, where the separation originated, and where it corresponds relatively well to reality, there is growing recognition of the need for integration.

Complex land husbandry systems are not a new invention. On the contrary, they were found to be the traditional practice in many of the hotter countries when Europeans first arrived. However, in the eyes of those with temperate experience, they seemed 'primitive', and efforts were made, often with success, to replace them by 'rational' systems in which the various forms of production were separated. It is only in recent years that the sound ecological basis of the old methods has been recognized, and that attempts have been made to develop them, and to accumulate and transmit information on them.

For its full effect, integrated land husbandry needs to be based on a correct estimation of the appropriate and sustainable use of each piece of land. This is unlikely to be achieved spontaneously, especially where there is pressure from graziers, since all land is potential grazing land in their eyes. Rather than rely on *laissez-faire*, it seems prudent to create an institutional framework for sound land-use, with suitability established and mapped by an independent body, and with incentives and disincentives to ensure proper use.

SELF-SUFFICIENCY: HAVEN OR TRAP?

It is almost universally agreed today that countries and even districts should aim at self-sufficiency in food. It has become the conventional wisdom to claim that people are hungry partly because, in the wake of colonialism, they devote precious land to the growing of cash crops for sale to rich countries instead of using it to feed themselves.

The popular support for self-sufficiency is contrary to the findings of economics. The principle of comparative advantage states that all societies will be better off in material terms if each specializes in what it produces relatively efficiently, using the surplus to pay for imports of what it produces relatively inefficiently. Where cash crops have a greater value than the staple food that could be produced on the same land, it is by growing them that maximum prosperity can be gained. If cash crops have failed to enrich those who grow them, it is not because they are cash crops but because the system of land ownership, marketing, and processing gives the profits to absentee landlords, middlemen, and the shareholders of multinational firms. What is needed is a change not in the crop but in the economic system. To call for peasants to return to self-sufficiency in staple foods is to condemn them permanently to subsistence farming.

The ecological basis for the relative efficiency of the hotter countries in growing cash crops lies in three facts. Firstly, their flora is far richer than that of the temperate zone, so that there are many valuable plants that can be grown only in hot countries, against only a few that flourish better in cool ones. Secondly, the characteristic tropical cash crops mostly grow on perennial plants, which can reach deep into the soil and whose high growth-rate ensures early maturity and fruition. Thirdly, most staple crops are annuals, and their widespread extension brings the risks of degradation and erosion discussed above at length.

Potentially the most valuable cash crop of all is timber. Investment in forest plantations is already very rewarding, with rates of return of 10–20 per cent per annum in many tropical countries, against 5 per cent or less in temperate ones. The advantage could well become even greater with research into improved plantation techniques, and with world timber prices seeming certain to rise in real terms. High growth-rates, and relatively low costs in terms of erosion, make it likely that the relative efficiency will lie permanently with hotter countries.

It is not suggested that the tropics should stop growing food and should turn all their land over to cash crops and timber: that would be a caricature of the argument. A balanced land husbandry will have plenty of room for annual plants alongside the permanent crops. But both ecologically and economically, the latter are likely to be the foundation of rural prosperity.

SOIL RENOVATION

With so many millions of hectares of land severely eroded and degraded, it will in many cases be necessary to renovate the soil before its productivity can be restored by complex husbandry systems. This can be achieved, except where hard rocks have been exposed, by using heavy rippers to break up compacted or indurated layers down to a depth of as much as 80 cm, thus creating channels through which rainwater can enter and gases can be exchanged, and reopening pores in which water can be stored. The product of this treatment may be only a pseudo-soil, lacking the organic matter and the living organisms of a true soil. These can be re-introduced either artificially or by allowing their slow development under cultivation.

Such operations are expensive, and it is only practical to undertake them once and for all. The renovated soil must therefore be securely protected against a renewal of degradation and erosion. In many cases this can be achieved simply by ensuring the adoption of systems that maintain a permanent cover of perennial plants. In other cases, and particularly on sloping land with an aggressive rainfall regime, erosion control methods need to be applied. The principal technique is the construction of terraces or of ditches and banks to conduct surplus water at safe speeds to specially reinforced evacuation channels.

OVERCOMING OBSTACLES

To introduce – or reintroduce – complex land husbandry systems, and to carry out thorough soil renovation, amounts in effect to a great upheaval in the way of life of a community. The changes may be far from trivial. For example, the intensive working of a wider range of crops is likely to mean alterations in the seasonal cycle of activity, and in the division of labour. The shift from extensive grazing to intensive animal husbandry will mean an increase in the work load and perhaps its transfer from child-shepherds to adults. The change from reliance on spontaneously grown fuel-wood, collected often by women, to the planting and tending of firewood trees, perhaps mainly by men, may mean an altered balance of roles for the sexes. There are many possible points of conflict with traditional values.

Rural development projects have in the past rarely been carried out with sufficient awareness of the offence that they may cause. The essential decisions have too often been made by technically qualified agronomists with no special understanding of social, psychological, or cultural questions. Programmes are initiated at the level of national government without adequate consultation of local people. The absence of effective democracy in many countries increases the danger that the views of those affected will not be heard, or that they will express themselves in more or less violent resistance.

The process of soil renovation is particularly likely to cause offence. It requires mechanical work of such magnitude that there is no practical alternative to the use of powerful machinery driven by highly skilled operators. The local community is thus inevitably the spectator of initial operations so massive that they destroy many familiar landmarks and boundaries. It is easy to see how the experience may cause reactions of rage or panic. Such projects are likely to be made acceptable only by very careful preparation, preferably involving visits to areas that have visibly benefited from recent treatment.

The spiniest problem of all is that of over-grazing. There are comparatively few purely grazier peoples, but they have a long history of bellicosity, and governments are wary of arousing their anger. On the other hand there must be hundreds of millions who own one or two animals for family use; although the productivity is low, such livestock is an important asset for its owners. Radical reform of the laws and taxes on grazing will be politically unpopular, unless very astutely allied with incentives and compensations. As yet, no government seems to have realized the size of the danger, or the importance of finding a solution.

REAPING THE REWARD

Some idea of the difference that massive investment could make is given by the joint study already referred to above (FAO 1982). Three sets of assumptions are made in predicting future food situations. The first supposes 'low inputs', that is to say a continuation of present crops and techniques. The third, 'high inputs', assumes a change to optimum mixtures of rainfed crops, with use of high-yielding cultivars, optimum fertilizer application, chemical pest, disease, and weed control, minimum fallow periods, complete conservation measures, general mechanization, and with sale of surpluses on world markets. The second set of assumptions, 'intermediate inputs', lies between the two extremes.

As mentioned above, with low inputs the excess population of critical areas is expected to rise from 278 to 503 million by the year 2000. With intermediate inputs it would fall to 141 million, and with high inputs to 48 million, of whom 42 million would be in South-West Asia (Afghanistan, Iran, Turkey, and the Arab world east of Suez).

From the standpoint of the present paper, this FAO study can be criticized on several counts. Firstly, it ignores comparative advantage and assumes a goal of food self-sufficiency for each country and region. Secondly, its high-input model is based on the application of temperate-style solutions from rich countries to the problems of poor countries with hot climates. Then, in spite of the participation of the International Institute for Applied Systems Analysis, the study considers agriculture in isolation from wider systems, assuming, for example, that the substitution of capital for labour is a good thing even where labour is by far the more abundant resource. As for forestry, it is not even considered, though the title of the study refers broadly to the 'population-supporting capacities of land', not merely to its food-producing possibilities.

In spite of such criticisms, the overall message is clear: with adequate investment, the productivity of land in the poorer countries could be very greatly increased. The number of people at risk of famine from a combination of crop failure and administrative breakdown could be reduced to a minimum by the year 2000. Such a situation might even be maintained through to the time when the human population at last ceases to increase, probably in the second half of the next century.

The benefits would be felt very much more widely than only in what are at present potential famine areas. Adequate quantities of food, timber and fuelwood, or charcoal would be available everywhere at reasonable prices. Intensive production on good land would make it possible to release large areas for conservation or recreation. Above all, the movement towards climax conditions over much of the world's land would make for a widespread improvement of the effective climate, with less extremes of temperature and humidity, and fewer dry winds and dust storms. Even the background climate could be expected to change, with increased evaporation resulting in improved rainfall.

COUNTING THE COST

There can be no question here of an exact estimate of the cost of a world-wide programme of soil renovation, erosion control, agricultural improvement, and forest plantation and management. An order of magnitude may nevertheless be indicated.

The most expensive single item would be soil renovation, in which there is no escape from the use of heavy machinery using great quantities of fuel. The most carefully costed study of such a programme is perhaps that which was carried out in Algeria between 1960 and 1965 (Stewart 1975). For the treatment of 13 million hectares it was estimated that the cost would be between 12 and 15 billion Ffr., depending on the degree of mechanization. Allowing for inflation and changes in parity, the corresponding figure today would be approximately $500 per hectare. Assuming that 500 million hectares of eroded land need such treatment in the hotter countries, the total would be of the order of $250 billion.

For the rest, the investment would be in the nature of an initial injection of resources to start the cycle of improvement. Once under way, increasing production should finance itself. A round figure of $100 billion will be assumed here, to pay for things like the cost of setting up crop-breeding stations, research centres, extension services, fertilizer factories, and tree nurseries. The total investment would thus be $350 billion, of the same order of magnitude as the value of one year's world trade in crude petroleum ($332 billion in 1980).

There would also be costs in terms of adjustments necessary in the world economy. At present the rich countries support farm incomes by keeping up food prices and subsidizing inputs of capital and non-renewable resources. Temperate agriculture thus produces surpluses which in resource terms are very expensive, but which are dumped on world markets, or given away, bypassing markets altogether. The effect is to depress the prices that farmers in the poorer countries can get for any surplus they produce. Moreover, inputs are excessively expensive for them, since they have to compete with buyers who are not only relatively rich but also subsidized.

Similarly, many temperate countries subsidize forestry or apply tariffs to processed-wood products, making it harder for the hotter countries to establish forest-based industries. This is wasteful in terms of world resource use, since it takes two or three times as much land to grow a given quantity of wood under temperate climates, and there is a high energy cost in transporting logs great distances before processing them.

Adjustments in farm and forest policy are needed if the rural people of the poorer countries – the majority of the world's population – are to improve their lot; but more important still is the matter of reducing the gap in income between rich and poor countries. Famine is simply an acute manifestation of the chronic poverty in which hundreds of millions of people live, a poverty which results from their inability to compete in world markets with the citizens of countries whose productivity is maximized by the possession of most of the world's capital. Some economists have suggested that economic growth could enable redistribution to be achieved by raising the level of income and consumption of the poor without lowering that of the rich (Cheney et al. 1974). It is plain, however, that if the poor are to consume a larger share of finite material resources, this can only be at the price of reducing the share of the rich.

The cost of eliminating famine is thus both investment in the soil and vegetation of the areas at risk, and a more general redistribution of wealth and income. It seems a fair price to pay for creating a world in which all can afford the food that they deserve.

BIBLIOGRAPHY

Apeldoorn, G. Jan van (1981). *Perspectives on famine in Nigeria*. Allen and Unwin, London.

Biswas, M. R. and Biswas, A. K. (1979). *Food, climate and man*. Wiley, New York.

Blakeslee, Leroy L. *et al.* (1978). *World food production, demand and trade*. Iowa State University Press.

Cépède, Michel *et al.* (1964). *Population and food*. Sheed and Ward, New York.

Charney, J. et al. (1975). Drought in the Sahara: a biogeophysical feedback mechanism. Science, 187, 434–5.

Cheney, H. et al. (1974). Redistribution with growth. Oxford University Press.

Cronon, W. (1983). Changes in the land. Hill and Wang, New York.

Crosson, P. R. and Frederick, K. D. (1978). The world food situation. Resources for the Future, Washington D.C.

Dar, A. (1983). Causes of famine–droughts. Swedish University of Agricultural Science, Uppsala.

FAO (Annual). Production yearbook. FAO [Food and Agriculture Organization], Rome.

FAO (1978). The state of natural resources and the human environment for food and agriculture. In State of food and agriculture 1977. FAO, Rome.

FAO (1982). Potential population supporting capacities of lands in the developing world, FPA/INT/513. FAO, Rome.

FAO (1983) (in progress). 1980 world census of agriculture. FAO, Rome.

Garcia, R. V. (1981). Drought and man. Pergamon Press, Oxford.

Gilland, B. (1979). The next seventy years. Abacus Press, Tunbridge Wells.

Hardin, G. (1986). Human ecology: the subversive conservative science. In Human ecology: a gathering of perspectives (ed. R. J. Borden). Society for Human Ecology, College Park, Maryland, USA.

Hardin, G. and Baden, J. (eds) (1977). Managing the commons. Freeman, San Francisco.

Hutchinson, J. (1969). Population and food supply. Cambridge University Press.

Klatzmann, J. (1975). Nourrir dix milliards d'Hommes. Presses Universitaires de France, Paris.

Kretchmer, N. (1972). Lactose and lactase. Scientific American, 227, 70–9.

McCracken, R. D. (1971). Lactose deficiency: an example of dietary evolution. Current Anthropology, 12, 479–517.

Ottermann, J. (1974). Baring high-albedo soils by over-grazing. Science, 186, 531–3.

Parikh, K. and Rabar, F. (eds) (1981). Food for all in a sustainable world. International Institute for Applied Systems Analysis, Laxenberg.

Pimental, D. et al. (1976). Land degradation: effects on food and energy resources. Science, 194, 149–55.

Revelle, R. (1976). The resources available for agriculture. Scientific American, 235, 165–78.

Sahlins, M. (1974). Stone age economics. Tavistock, London.

Steele, F. and Bourne, A. (1975). The man/food equation. Academic Press, London.

Sinclair, A. R. E. and Fryxell, J. M. (1985). The Sahel of Africa: ecology of a disaster. Canadian Journal of Zoology, 63, 987–94.

Stewart, P. J. (1975). Fight erosion or emigrate, Algerian peasantry at the crossroads, Discussion Paper, No. 69. Institute of Development Studies, University of Sussex.

Stewart, P. J. (1979). Islamic law as a factor in grazing management. Commonwealth Forestry Review, 58, 27–31.

Stewart, P. J. (1981). Forestry, agriculture and land husbandry. Commonwealth Forestry Review, 60, 29–34.

Valdès, A. (1981). Food security for developing countries. Westview Press, Boulder, Col.

INDEX